International Series in Operations Research & Management Science

Volume 183

For further volumes:
http://www.springer.com/series/6161

Jeffrey W. Herrmann
Editor

Handbook of Operations Research for Homeland Security

Springer

Editor
Jeffrey W. Herrmann
A. James Clark School of Engineering
University of Maryland
College Park, Maryland, USA

ISSN 0884-8289
ISBN 978-1-4614-5277-5 ISBN 978-1-4614-5278-2 (eBook)
DOI 10.1007/978-1-4614-5278-2
Springer New York Heidelberg Dordrecht London

Library of Congress Control Number: 2012948967

Printed on acid-free paper

Springer is part of Springer Science+Business Media (www.springer.com)

This book is dedicated to my family, whose love and encouragement are the most wonderful of the many blessings that God has given to me.

Preface

The purpose of this book is to enlighten policy makers and decision makers about the power of operations research (OR) to help organizations plan for and respond to terrorist attacks, natural disasters, and public health emergencies.

The intended audience includes (1) policy makers and decision makers in federal, state, and local government agencies related to homeland security, emergency management, and public health preparedness; (2) nongovernment organizations with similar missions; and (3) operations research workers, students, and scholars.

This book begins with a general overview of how operations research can be used and then provides specific examples of operations research techniques and their application to homeland security problems.

Chapter 1, "Using Operations Research Methods for Homeland Security Problems," describes OR techniques and highlights their use to solve homeland security problems. This is intended to give policy makers and decision makers insight into what OR can do and how it is relevant to their particular concerns and problems. After discussing how to use OR, the chapter reviews four areas: (1) understanding what has happened; (2) considering what could happen; (3) deciding what to do; and (4) finding the best solution.

Chapter 2, "Operations Research and Homeland Security: Overview and Case Study of Pandemic Influenza," provides additional context by describing the origins of operations research (OR) at the beginning of World War II and reviewing OR tools and techniques that can be used for three types of decisions: long range (strategic), medium range (tactical), and real time (operational). It then discusses the results of a study of the process of manufacturing, delivering, and administering the flu vaccine during the 2009 influenza pandemic.

Chapter 3, "Deployed Security Games for Patrol Planning," presents models that can help security forces generate randomized security policies that are more difficult for adversaries to predict and exploit. The chapter describes a generic mathematical formulation of these models, presents some of the results that have allowed these systems to be deployed in practice, and outlines remaining future challenges. The chapter discusses the deployment of these systems in two homeland security applications: (1) the police at the Los Angeles International Airport use these models to randomize the placement of checkpoints on roads entering the airport and the routes of canine unit patrols within the airport terminals and (2) the Federal Air Marshal Service uses these models to randomize the schedules of air marshals on international flights.

Chapter 4, "Interdiction Models and Applications," describes an approach for assessing the vulnerabilities of operational systems. Interdiction models can be used to answer the following questions: How is the system operated? What are the vulnerabilities of that system? How can we invest to make the system more resilient? This chapter discusses the application of these models to four problems: (1) delaying an enemy's development of a nuclear weapon; (2) understanding the vulnerabilities of an electric power system; (3) locating sensors that can rapidly detect the contamination of a municipal water system; and (4) locating radiation sensors to detect nuclear smugglers. The chapter also discusses practical implications and insights obtained from the models in these applications.

Chapter 5, "Time Discrepant Shipments in Manifest Data," presents an innovative data mining technique for identifying suspicious activity by identifying and recording unusual patterns that have high activity. The chapter focuses on container shipments to US ports. By analyzing the origin–destination pairs, container contents, and shipment dates, the procedure creates a variety of graphs to visualize the shipment activity, measures the discrepancy of shipment patterns, and distills a data fossil that aggregates these patterns over time.

Chapter 6, "Achieving Realistic Levels of Defensive Hedging Based on Non-Monotonic and Multi-Attribute Terrorist Utility Functions," addresses the problem of allocating limited resources to defend a set of targets. When there is uncertainty about which targets the terrorists are most likely to attack, decision makers are likely to insist on some degree of "hedging" (defending targets with only moderate value). The work discussed in this chapter uses game theory to find the optimal strategy for the defender and shows that non-monotonic attacker objective functions do typically yield greater hedging.

Chapter 7, "Mitigating the Risk of an Anthrax Attack with Medical Countermeasures," presents a simulation model that can be used to prepare for a bioterrorism attack that releases anthrax spores and exposes thousands of persons to this deadly disease. The model predicts the expected number of deaths using information about the size of the population, the number exposed, the progress of the disease, the resources available for distributing medication and treating the ill, and the size of local medication stockpiles. The chapter also presents a risk management approach for allocating a limited medication stockpile to multiple cities to minimize the expected number of deaths. The results show that the optimal allocation can be quite different from allocations that are proportional to population size.

Chapter 8, "Service Networks for Public Health and Medical Preparedness: Medical Countermeasures Dispensing and Large-scale Disaster Relief Efforts," discusses the results of collaborations between OR experts, the Centers for Disease Control and Prevention, and other public health agencies to develop useful tools for planning efforts to respond to public health emergencies. Such emergencies include bioterrorism attacks, naturally occurring pandemics, and severe meteorological and geological events. In paricular, the chapter describes OR models for optimizing mass dispensing operations.

Chapter 9, "Disaster Response Planning in the Private Sector and the Role of Operations Research," discusses how organizations in the private sector can be effective first responders in the aftermath of disasters. The chapter describes the disaster response planning process implemented by The Home Depot, discusses the role of OR methods to assist decision making, and presents an optimization model to improve advance purchasing and inventory allocation.

This book is the result of a team effort. The authors who contributed their valuable time to produce the informative chapters that comprise this volume must be thanked first. I appreciate their effort to develop, write, and revise their chapters and complete the necessary supporting tasks.

I would like to thank Fred Hillier for inviting me to edit this handbook, for it is an honor to be a part of this distinguished series. My thanks also go to Matthew Amboy, who guided the process of transforming our chapters and ideas into a book.

In addition to caring for our family, my wife Laury provided useful editorial assistance, and I appreciate the time and effort that she spent helping me on this project.

Finally, I am indebted to my generous family and the wonderful friends, colleagues, teachers, and students who have shared their wisdom, energy, and talents with me.

Maryland, USA Jeffrey W. Herrmann

Contents

Contributors

James Abello DIMACS Center, Rutgers University, Piscataway, NJ, USA

Bernard Benecke Global Disease Detection and Emergency Response, Center for Global Health, Centers for Disease Control and Prevention, Atlanta, GA, USA

Vicki Marion Bier Department of Industrial and Systems Engineering, University of Wisconsin-Madison, Madison, WI, USA

Jaime Marie Bonorato University of Wisconsin, Madison, WI, USA

Mikey Chen DIMACS Center, Rutgers University, Piscataway, NJ, USA

Nedialko B. Dimitrov Operations Research Department, Naval Postgraduate School, Monterey, CA, USA

Matthew J. Drake Palumbo and Donahue Schools of Business, Duquesne University, Pittsburgh, PA, USA

Özlem Ergun School of Industrial and Systems Engineering, Georgia Institute of Technology, Atlanta, GA, USA

Stan Finkelstein Massachusetts Institute of Technology, Cambridge, MA, USA

Jeffrey W. Herrmann A. James Clark School of Engineering, University of Maryland, College Park, MD, USA

Manish Jain Computer Science Department, University of Southern California, Los Angeles, CA, USA

Juan F. Jara Industrial Engineering Department, University of Chile, Santiago, Chile

Gonca Karakus School of Industrial and Systems Engineering, Georgia Institute of Technology, Atlanta, GA, USA

Paul Kerl School of Industrial and Systems Engineering, Georgia Institute of Technology, Atlanta, GA, USA

Pinar Keskinocak School of Industrial and Systems Engineering, Georgia Institute of Technology, Atlanta, GA, USA

Christopher Kiekintveld Computer Science Department, University of Texas, El Paso, TX, USA

Richard C. Larson Massachusetts Institute of Technology, Cambridge, MA, USA

Eva K. Lee Center for Operations Research in Medicine and HealthCare, School of Industrial and Systems Engineering, NSF I/UCRC Center for Health Organization Transformation, Georgia Institute of Technology, Atlanta, GA, USA

David P. Morton Graduate Program in Operations Research, The University of Texas at Austin, Austin, TX, USA

Fernando Ordóñez Industrial Engineering Department, University of Chile, Santiago, Chile

Neel Parikh DIMACS Center, Rutgers University, Piscataway, NJ, USA

Ferdinand Pietz Strategic National Stockpile, Office for Public Health Preparedness Response, Centers for Disease Control and Prevention, Atlanta, GA, USA

Julie L. Swann School of Industrial and Systems Engineering, Georgia Institute of Technology, Atlanta, GA, USA

Milind Tambe Computer Science Department, University of Southern California, Los Angeles, CA, USA

Anna Teytelman Massachusetts Institute of Technology, Cambridge, MA, USA

Jason Tsai Computer Science Department, University of Southern California, Los Angeles, CA, USA

Monica Villarreal School of Industrial and Systems Engineering, Georgia Institute of Technology, Atlanta, GA, USA

Chen Wang University of Wisconsin, Madison, WI, USA

Chapter 1
Using Operations Research Methods for Homeland Security Problems

Jeffrey W. Herrmann

Abstract This chapter describes operations research (OR) techniques and highlights their use to solve homeland security problems. This is intended to give policy makers and decision makers insight into what OR can do and how it is relevant to their particular concerns and problems. After discussing how to use OR, the chapter reviews four areas: (1) understanding what has happened; (2) considering what could happen; (3) deciding what to do; and (4) finding the best solution. The techniques available for understanding what has happened include data mining and monitoring time series data. Considering what could happen requires risk analysis and using queueing models and simulation models. Decision analysis and game theory can be used to make decisions, but finding the best solution requires formulating and solving an optimization problem. Because the examples illustrate how specific techniques have been applied, homeland security professionals can use this chapter to understand what operations research can offer and identify the techniques that are relevant to their problems. The chapter also includes references to essential books and articles that can provide more information.

1.1 Introduction

Operations research (OR) is "the discipline of applying advanced analytical methods to help make better decisions" (INFORMS 2011). OR methods provide understanding into what has happened in the past and what could happen in the future. As the OR community increased its consideration of the problems that occur in the area of homeland security, surveys of this growing body of knowledge also appeared. Different surveys took quite different approaches to organizing previous work.

J.W. Herrmann (✉)
A. James Clark School of Engineering, University of Maryland, College Park, MD 20742, USA
e-mail: jwh2@umd.edu

J.W. Herrmann (ed.), *Handbook of Operations Research for Homeland Security*, International Series in Operations Research & Management Science 183, DOI 10.1007/978-1-4614-5278-2_1,

The survey by Wright et al. (2006) adopted a framework that followed the science and technology portfolios that the U.S. Department of Homeland Security had at that time. [These portfolios have since become a similar list of focus areas (DHS 2011).] The Wright survey reviewed OR related to countermeasures, border and transportation security (including border security, airline security, port and rail security, and truck security), critical infrastructure protection, cyber security, emergency preparedness and response, and threat analysis. The discussion of emergency preparedness and response research covered work on emergency response, including the studies of the operations of the New York City fire and police departments in the 1970s. A recurring problem in this area is one of deciding where to locate resources such as fire companies, emergency warning sirens, and cellular base stations. The Wright survey also covered work on modeling evacuations from buildings and from cities and problems related to responding to disasters. The study discussed how OR work has covered problems in three of the four phases of emergency management: planning for disasters, preventing disasters, and responding to disasters. The authors found no papers on problems related to recovering from disasters.

The review by Altay and Green (2006) focused on OR related to disaster operations management. Like the Wright survey, the Altay and Green review considered the four phases of emergency management: mitigation (preventing or reducing the impact of disasters), preparedness (planning the response), response, and recovery. The Altay and Green review systematically identified 109 articles describing OR work in these areas by searching research databases for papers published after 1980 and examining the citations of these articles. They found that 44% of the OR papers were related to mitigation, 21% were related to preparedness, 24% were related to response, and 11% were related to recovery. They also described the distribution of these papers by the disaster type, the decade published, the methodology used, and the type of research [management science, management engineering, or management consulting (cf. Denizel et al. 2003)]. The review showed which papers fell into which categories, but did not explain what each paper did. The survey concluded by emphasizing the need for more research, particularly research that considers the coordination of multiple agencies, research that uses soft methodologies, research that considers the recovery phase, research related to business continuity, and research into infrastructure design.

Because these two surveys have provided a good overview of the range of problems to which OR has been applied, this chapter focuses instead on describing OR techniques and highlighting their use to solve problems related to homeland security. This is intended to give policy makers and decision makers insight into what OR can do and how it is relevant to their particular concerns and problems.

After discussing how to use OR, the chapter reviews four areas (1) understanding what has happened (through data mining and monitoring time series data), (2) considering what could happen (using risk analysis, queueing models, and simulation models), (3) deciding what to do (with help from decision analysis and game theory), and (4) finding the best solution (by solving mathematical programs, facility location problems, and vehicle routing problems).

1.2 How to Use Operations Research

The examples discussed in this chapter demonstrate numerous ways to apply OR methods successfully to understand problems and make better decisions. There is no single way to proceed with OR methods. Within this diversity, a key process for using OR begins with creating a model that represents a real-world situation, analyzing or executing that model, and then translating the results into something that has meaning in the real-world situation. The following steps illustrate how this can be done:

1. Describe the problem.
2. Design a model and collect relevant data.
3. Build and validate the model.
4. Use the model to gain insight into the problem.
5. Translate the results into a solution.
6. Implement and evaluate the solution.

Ideally OR is a process that iterates through the phases of description, explanation, and testing (Meredith 2001). Krings and Azadmanesh (2005) described OR as a "transformation" process. This process begins by transforming the problem into a model and mapping data about the problem into the parameters of the model. The model is classified, analyzed, and optimized. The answer that results must be translated by reversing the mappings used to create and instantiate the model, which yields a solution that can be implemented in the real-world application.

OR has even greater benefits for organizations that go beyond using OR for a single study (or decision-making setting) and institutionalize OR. The early OR groups successfully improved U.S. military operations during World War II through their close relationship and alignment with the military decision makers (Defense Science Board 2009). Establishing an effective OR capability requires (1) maintaining leadership commitment and support for OR professionals and their analysis, (2) establishing OR as a distinct career field in which professionals are recruited and developed, (3) using OR across a range of operations and settings, and (4) aligning OR analysts, decision makers, and operations personnel (Defense Science Board 2009).

1.3 What Happened? Finding Patterns in Data

In the quest for understanding, a natural starting point is to consider what has happened in the past. Organizations that have large collections of data about persons and their transactions (like customers or travelers and their purchases or phone calls) have the ability to run standard reports on these data sets to monitor their operations and support routine decision making. In addition, users may submit queries to obtain data for answering questions that they develop. Finding patterns that are not even imagined, however, require more sophisticated techniques.

1.3.1 Data Mining

Data mining, unlike the queries of curious analysts, is a knowledge discovery technique that automatically discovers relationships that were previously unknown.

In general, data mining starts with collecting and managing the data that are relevant to the problem. Data mining techniques can consider the data in multiple, heterogeneous databases as well as very large transactional databases. The analyst may specify a list of variables (fields) in that data to consider. Limiting the procedure to certain fields avoids wasting time looking for relationships involving irrelevant data. Data mining algorithms then begin processing the data to identify new relationships among the variables. There are many different analysis techniques. For descriptions of the key algorithms used in data mining, see, for example, Hastie et al. (2009).

The problem of identifying terrorists before they attack is a natural domain for knowledge discovery in general and data mining in particular. Algorithms that can search multiple databases to find suspicious actions and lead officials to the terrorists would be extremely valuable. The efforts to do this have led to concerns about privacy and to assessments that data mining is not the appropriate tool because the rate of false alarms (likely to be high) will waste valuable investigative resources and violate civil liberties (Jonas and Harper 2006). Moreover, data mining relies upon finding patterns that occur repeatedly, but terrorist cells are unique and can deliberately act in ways unlike past groups.

Data mining has other homeland security applications, however. In 2004, the U.S. Government Accountability Office (GAO) reported that government agencies were using or planning at least 29 data mining projects for analyzing intelligence and detecting criminal and terrorist activities (Fienberg 2008).

Data mining can also be used to understand the activities of terrorist organizations and extremist groups. A study of the web sites belonging to such organizations in the USA, Latin America, and the Middle East revealed that the groups from the Middle East have most actively used the Internet. Their web sites were technically sophisticated and provided the richest multimedia content. The U.S. groups used Internet communication technologies, including chat rooms and electronic commerce. The web sites of Latin American groups were not as sophisticated technically and used Internet communication technologies less effectively (Qin et al. 2008).

Data mining has been used to develop intrusion detection systems that can protect computer networks by identifying unauthorized or abnormal activities that may be cyber attacks (Chen et al. 2005). Chen et al. started with a dataset generated by a network security module that logged all system calls over 35 days, converted the raw data into human readable ASCII form, and then extracted individual sessions, processes, and system calls. Data about normal sessions and attack sessions from 5 days were treated as the training data set. (Data about normal sessions and attack sessions from a different 5-day period were treated as the testing data set.) Chen et al. used the training data set to train an artificial neural network that could determine whether a process was abnormal based on information about the system calls in that process. They also used the training set to train a support vector machine that can perform the same type of function. Essentially, the artificial neural network (ANN) and the support vector machine (SVM) are sets of rules that recognize deviations from normal usage patterns and flag them as intrusions. Although based on the same training data, the two models use different types of rules and thus sometimes disagree. Chen et al. then tested the two models on the testing data set. Because the researchers knew which processes and sessions were normal and which were attacks, they could determine that the SVM could detect all of the attacks with a false-positive rate (the fraction of normal sessions classified as attacks) of 8.5%. Although the ANN could also detect all of the attacks, its false-positive rate was 39.3%. From this and other similar results, they concluded that the SVM performance was superior to the ANN performance.

1.3.2 Monitoring Time Series Data

Monitoring variables that change from day to day can provide important, timely information about what has happened. Those who monitor a system or population are most interested in when something has changed. This change (e.g., the failure of a component in a machine or the release of a pathogen in a population) will affect the distributions of the variables that are being monitored. A variety of statistical techniques have been developed to analyze the observed values and signal when a significant change has occurred.

In the context of homeland security, syndromic surveillance is a well-known effort to observe and analyze time series data. Syndromic surveillance focuses on data about the symptoms of individuals who seek care in emergency rooms, doctors' offices, and other health care services. The goal of syndromic surveillance is to detect changes that signal the outbreak of an infectious disease or bioterrorism attack.

Many of the analytical techniques used in syndromic surveillance are derived from the methods used in statistical process control, which was developed for monitoring a manufacturing process and determining when it changes from normal (in control) to out of control. The key planning steps are to develop a model of how the process normally behaves and to develop a rule for sounding an alarm. The rule usually corresponds to the occurrence of an event that should be rare when the process is in control but is more likely when the process is out of control. Monitoring the process requires collecting data about the process, calculating the relevant statistic at regular intervals, and then using the rule to determine whether an alarm should be sounded.

The simplest algorithms work on only a single variable, like the sample mean of some recent observations. Shewhart's algorithm and the cumulative sum (CuSum) algorithm are common and can be extended to multivariate algorithms. Stoto et al. (2006) conducted experiments on these methods with data from emergency room admissions in Washington, D.C., and concluded that the CuSum algorithms are preferable to Shewhart's algorithm for syndromic surveillance.

Algorithms like Shewhart's algorithm raise an alarm by considering only the current value of the statistic, which makes the algorithm responsive to large changes but less useful for detecting changes that occur more slowly. Statistics that use more past data include the CuSum and the exponentially weighted moving average (EWMA). The EWMA, unlike the CuSum, gives more recent observations higher weights. Other useful time series models include the autoregressive moving average (ARMA) models, which add predictors to the model, which are useful in biosurveillance for handling the impact of weekends and holidays on the data. For more on these techniques, see Shmueli and Fienberg (2006) for their application to biosurveillance and Box et al. (1994) for the mathematical details.

Both EWMA and ARMA techniques have been used in ESSENCE II, a widely used syndromic surveillance system that systematically collects nontraditional, nonspecific indicators of health status, grouped into health syndromes among a patient population (Lombardo et al. 2004). The system collects information from hospitals, sentinel providers, over-the-counter pharmaceutical sales, and school-based absenteeism reports. It applies statistical algorithms to detect unexpected changes in the data and provides the information to health officials in a web-based application. To avoid the considerable effort required to train such algorithms, especially when multiple symptoms at multiple hospitals will be monitored, Sparks et al. (2010) developed adaptive EWMA and adaptive CuSum approaches.

Although syndromic surveillance is the most widely known application of time series monitoring in the context of homeland security, similar techniques could be used to monitor computer system attacks and other phenomena.

1.4 What Could Happen? Building Stochastic Models

Understanding what could happen often requires considering the uncertainty of future events. By far the most common way to handle an uncertainty is to describe it using a probability distribution. That is, it becomes a random variable. If one has a model of the system (or process) being studied, using random variables as the inputs to the model yields outputs of the system (such as measures of system performance) that are also random.

Stochastic models are models that have random variables. Stochastic models can be used to identify the distribution of possible outcomes of a process (as in risk analysis) and to estimate the performance of a system in which the events are random (for example, queueing models).

If the system is relatively simple and the probability distributions describing the random variables have convenient shapes, then mathematical analysis can be used to predict system performance exactly. Otherwise, mathematical equations can be used to estimate the system performance, or one can resort to detailed simulation models that sample the random variables and calculate the associated system performance.

1.4.1 Risk Analysis

Risk analysis considers the range of possible outcomes from a process and the likelihood and consequences of each outcome. The primary components of risk analysis are risk assessment, risk management, and risk communication (Modarres 2006). Risk assessment estimates likelihood and magnitude of consequences (usually losses) by considering what can go wrong, how likely is it, and what the losses (consequences) would be. Risk management is a decision-making process in which an organization tries to minimize and control risks by preventing problems, minimizing the consequences of problems (perhaps through buffers), and creating contingency plans.

Many homeland security and emergency preparedness activities are risk analysis, and a variety of tools have been developed for different types of risk assessment (Moore et al. 2010). Operations research techniques can improve the processes of risk analysis in various ways, as the following examples demonstrate. The following paragraphs focus on risk assessment, as the techniques for risk minimization fall generally into the area of optimization; see, for example, Jacobson et al. (2003) and Kobza and Jacobson (1996, 1997).

Terrorism risk is often modeled as the product of three components (Ezell et al. 2010): the threat (the probability of an attack), the vulnerability (the probability of an attack's success given that it occurs), and the consequences (the fatalities, injuries, economic impacts, and other losses that result from a successful attack). Adopting this view leads to a probabilistic risk analysis (PRA). Li et al. (2009) described the use of PRA to assess and rank the risks from natural hazards, human-induced accidents, and malicious acts in a community.

Designing an effective access control system requires information about the likelihood of threats and the probability of false alarms. Jacobson et al. (2000) used probability theory to develop an approach for estimating these parameters from alarm information generated about items that are known threats, items that are known not be threats, and other items. Their approach calculated both estimators and confidence intervals.

More generally, counterterrorism efforts assess the threats posed by terrorist organizations and the risk to particular facilities. The usefulness of these efforts can be improved by integrating them in a venue-specific risk assessment model (Shahar 2008). In this approach, the likelihood of an attack on a facility depends upon both the characteristics of the terrorists and the characteristics of the facility because the terrorists' capabilities and goals vary. For example, some terrorists want to maximize casualties, while others target high-profile facilities that require extensive planning.

The approach presented by Shahar starts by determining which terrorist organizations pose the greatest threat and should be included in the analysis. Then, the vulnerabilities of the facility are determined and a list of potential scenarios (types of attacks) is generated and evaluated. The last step integrates these results to estimate the likelihood that each terrorist organization will execute each attack type at that facility.

The use of PRA for terrorism risk assessment in particular has been criticized, however. Brown and Cox (2011) argue that "attack probabilities depend on what the attacker knows or believes, rather than on what the defender knows or believes. In contrast to risk analysis for defense against random events, risk analysis for reasoning attackers should consider how what the attacker discovers in the future may affect his future decisions." That is, when conducting risk assessments, it is important to consider the difference between random processes (like weather) that essentially ignore emergency and defensive preparations and intelligent adversaries (like terrorists) who look for the best way to attack. Probabilistic risk analysis techniques are more relevant in the first case, while it is important to use game theory models in the second.

Game theory can improve our understanding of "the nature of the key decisions that intelligent attackers and defenders must make" and emphasizes "that vulnerability and consequence are usually functions of the allocation decisions made by the players, not exogenous numbers or random variables" (Cox 2009). Cox also provided some simple examples illustrating the use of game theory in risk analysis.

Examples of using game theory to optimize defensive resource allocations will be discussed later in this chapter.

1.4.2 Queueing Models

Queueing systems refer generally to those systems in which various entities enter a system, wait for service by resources, receive service, and then depart. In the domain of homeland security, typical queueing systems include travelers waiting at security checkpoints and residents waiting to receive medication. In these queues, the customers are physically waiting in a line. On the other hand, some queues, like terror plots waiting to be investigated, do not appear as lines.

When studying queueing systems, it is important to predict their performance. Common performance measures include the average time that entities spend in the system, the average number of entities in the system, and the maximal rate at which the system can service entities. Queueing theory is concerned with developing mathematical models that can make these predictions (Hall 1991; Newell 1982). In some cases, the models can provide an exact answer; in others, the models provide only estimates. More generally, queueing theory provides guidance on how much capacity is needed to provide adequate service.

To design a passenger security screening process at a new airport terminal, Gilliam (1979) used queueing theory to determine the number of screening stations required and to determine how system performance degraded if the screening rate (the number of passengers screened per minute) decreased.

Points of dispensing (PODs), also known as mass dispensing and vaccination clinics, are a key component of the plans for responding to a bioterrorism attack. Public health staff would vaccinate residents of the area affected (for smallpox, for instance) or dispense medication to them (for anthrax, for instance). PODs are also used for influenza vaccination campaigns (like those opened in 2009 to provide vaccinations during the H1N1 influenza pandemic).

Carefully planning PODs is important. The health department must train the right number of people beforehand (although they can do some training at the time of need) and must assign the right number of workers to various roles when the clinic begins operations. They must consider the capacity of each clinic (the number of residents it can serve per hour) and how much time residents would spend in the clinic (the time in system, the flow time, or the throughput time). Clinic capacity affects the number of clinics needed and the total time needed to vaccinate the affected population. The time in system affects the number of residents who would be inside the clinic waiting for treatment; too many residents in the clinic (and not enough space) can cause crowding and confusion.

Queueing network models can be used to create easy-to-use modeling tools, implemented using spreadsheet software, to help public health officials plan PODs (Aaby et al. 2006a, b; Herrmann 2008). The models require information about the number of persons to be served, the stations that the persons will visit, and the processing times at each station. The model requires information about the variance of the interarrival times and the processing times. The models calculate the minimum number of staff required and, for a given staffing plan, estimate the POD capacity and how long persons spend in the POD.

Understanding how people will evacuate an area in an emergency is an important part of planning for emergencies such as hurricanes or fires in large buildings. A variety of simulation and analytical queueing models have been developed for this problem (Bakuli and Smith 1996; Wright et al. 2006).

Analyzing a particular building starts with modeling it as a queueing network. The nodes of this network are the channels through which evacuees must pass (e.g., hallways and stairwells). Some channels will have finite capacity; that is, there is a limit on how many people can pass through per minute. Each node is modeled as a queueing system. Given information about the rate at which evacuees arrive and the capacity of the channel, one can estimate the delays that occur and how long it will take for evacuees to leave the building. Bakuli and Smith (1996) used a queueing network model to quantify how increasing corridor width increases corridor capacity.

Queueing theory can also be used to understand how intelligence agents intercept and disrupt terror plots (Kaplan 2010). In Kaplan's model, the customers that move through the queueing system are terror plots, and the servers are intelligence agents. The queue consists of terror plots that are in the planning stage. If the planning is completed before the terror plot can be served (intercepted) by an intelligence agent, then it is successful; otherwise, it is intercepted and disrupted. Kaplan presented and analyzed a mathematical model of this queueing system, in which the arrival of terror plots, the time required to plan them, and the time needed to intercept and disrupt them are all random variables. The model was used to estimate the rate at which terror plots are intercepted and disrupted (for a given number of intelligence agents) and how changing the number of intelligence agents would change this rate. The model was also used to estimate the number of terror plots in the planning stage. Of course, the values required to populate the model are sensitive, so Kaplan presented only hypothetical examples, but security agencies like the FBI, MI5, and the Shabak can generate the numbers for their operations.

1.4.3 Simulation Models

Simulation models are computer programs that replicate the performance of a system. State-of-the-art simulation software allows one to construct and run a wide variety of simulation models with a large number of objects in them. The simulation software can calculate statistics about system performance measures, dynamically monitor the value of variables during a simulation run, and show the behavior of the system using high-quality animations. Simulation can be used for almost any type of system, from very small to very large, and can simulate a long time period within seconds on a computer. Simulation models can include great detail if one wants to include it, and can also ignore details in order to simplify the model and reduce computational effort. Within operations research, discrete-event simulation is the most common type of simulation, for it allows one to model the discrete entities that move through a system (ships, airplanes, trucks, cars, passengers, and bags, for example) and the discrete state of the system resources and entities (whether a machine is busy or idle or broken, or the condition of a sick patient, for instance). A popular text on simulation modeling is Law (2007).

Generally, modeling and simulation are viewed as an important technique for both planning and optimizing the responses to possible emergencies and training personnel to perform their roles, (O'Hara et al. 2010). A federated simulation architecture (based on the ideas of Jain et al. (2007)) that integrated multiple simulation models was used successfully to support a multi-departmental exercise in which participants exchanged information in a counterterrorism scenario (McCormick et al. 2010).

Simulation models can be used to represent large populations in order to predict how a disease spreads through a community. Das et al. (2008) presented a large-scale simulation model of how influenza would affect a large urban area (as part of an influenza pandemic). The model simulates the spread of the disease in the presence of mitigation strategies that combine vaccination, prophylaxis, hospitalization, and social distancing. The simulation predicts the total number infected, the number of deaths, the number denied hospital admission, the number denied vaccines and antiviral drugs, and an aggregate cost measure that combines healthcare cost and lost wages. The model is location specific and requires information about demographic and community features and the activities of the residents. Das et al. showed how the simulation model can be used to determine which controllable factors have a significant impact on the number who will become ill.

Aleman et al. (2011) described an agent-based simulation model of an influenza outbreak in greater Toronto (which has a population of almost five million people). The model uses information about each person's age, health, vaccination status, home location, work location, transportation route, and household membership to determine the probabilities that they become infected and then recover. By using a parallel computing cluster, their implementation of the model can run 1,000 60-day runs in 13 min. The model is used to evaluate the impact of mitigation strategies on the number of people infected. Their results showed the extent to which an outbreak is less severe as more infected persons stay home.

EpiSims, another large-scale agent-based simulation model, has been used to evaluate the impact of mitigation strategies on the transmission of smallpox (Michalak and Wilson 2006) and influenza (Stroud et al. 2007). The model includes details of the activities that persons perform each day and the different modes of transportation that they use. The model can evaluate school closures, household quarantine, administering antivirals, vaccination campaigns, social distancing, and other strategies.

Simulation is also commonly used to model specific operations, such as a point of dispensing, a security checkpoint, or a hospital. Miller et al. (2006) used a discrete-event simulation of a health care network (clinics, hospitals, and other facilities) to predict the resources required to treat the ill during a smallpox outbreak. Patients are diagnosed and then treated in intensive care units and other facilities. Their treatment requirements are determined randomly. Patients moved from one facility to another if they needed resources that are not available where they are. The model evaluates the utilization of resources in the health care network. Miller et al. used the model to determine the possible benefits of vaccination and quarantine in San Antonio, Texas.

1.5 What Should We Do? Making Decisions

Understanding should lead to better decision making. Although decision making is sometimes easy, it can be difficult when the decision requires considering many alternatives, many interacting issues, multiple criteria, or uncertainty in the outcomes. Moreover, humans often make poor decisions, even after spending significant time and effort on the process.

Decision making is a process that involves identifying the most important objectives, defining the criteria that will be used to screen and sort alternatives, identifying alternatives, evaluating the alternatives on the key criteria, and selecting the best alternative. Models that provide insight into what has happened and models that predict what could happen yield valuable information for decision makers.

When the number of alternatives is large, optimization models provide a way to search the space of possible solutions and find superior solutions (as discussed in the next section).

This section covers decision analysis, which includes some useful tools for performing these steps, and game theory, which provides insights into making decisions that involve other rational actors.

1.5.1 Decision Analysis

Decision analysis "provides structure and guidance for thinking systematically about hard decisions" (Clemen and Reilly 2001). As a body of knowledge, decision analysis is generally prescriptive because it suggests how one should make a decision. This begins with presenting a standard decision analysis process.

The techniques typically associated with decision analysis include structuring tools like influence diagrams and decision trees, using utility functions to model preferences about multiple criteria and uncertainty, and estimating the value of information. The textbook by Clemen and Reilly (2001) is a good introduction to these techniques.

An influence diagram is a model of a decision that represents the relationships between decisions, uncertainties (if there are any), and the objectives. An influence diagram has arcs and nodes. The nodes are decision nodes, chance nodes, intermediate consequences nodes, and a payoff node. The arcs can represent relevance or sequence relationships. Arcs into a chance node represent relevance, meaning that something affects that chance (the probabilities associated with that chance). (For instance, weather affects the chance of an accident; or the speed at which one drives affects the chance of an accident.) Relevance arcs can also go into consequences and payoffs, obviously, since the decisions and outcomes affect these. Arcs into a decision node indicate sequence. That is, they show what is known when the decision must be made. For instance, a driver observes the weather before he decides his speed.

For example, Paté-Cornell (2009) presented an influence diagram that showed how the consequences of a terrorist attack depend upon aspects of the terrorists (their preferences, their supply chain, and whether Americans assist them). These factors influence the terrorists' choice of target, weapon, and means of delivery. All of these factors influence the intelligence signals collected and analyzed by those protecting the USA. These signals and analysis, in turn, influence the countermeasures selected by American decision makers. The terrorists' choices and the countermeasures selected affect the outcome (the potential consequences of an attack). The influence diagram is useful both as a way to understand and communicate how different factors affect the outcome and also as a quantitative model for estimating the probability of an attack. The model was later extended to consider the terrorists' decisions and objectives (Paté-Cornell and Guikema 2002).

A decision tree displays additional detail about a decision by explicitly showing the alternatives available to a decision maker and the possible outcomes of random events. Decision trees can be used to identify the best alternative at any point in time. They also help stakeholders understand the range of possible outcomes, along with the likelihood of each outcome.

For example, Martonosi and Barnett (2006) used a decision tree to model an airline passenger screening process in which some passengers are selected for secondary screening while most go through primary screening. The model was used to calculate the probability that a terrorist with weapons will be able to board an aircraft. This probability was based on the probability that the terrorist is classified as high risk, the probability that low-risk passengers are selected for secondary screening, and the effectiveness of the screening procedures to detect weapons. Thus, the model could be used to make some general points about which improvements in screening effectiveness have the most impact.

When a decision requires evaluating and making tradeoffs between multiple objectives, such as cost, performance, and time, it is often useful to have some way to combine the multiple objectives (the attributes) into a single measure that can be used to sort the alternatives and identify the best ones. A multiattribute utility (MAU) model does this by combining utility functions for each attribute to get an overall utility for each alternative. Each attribute's utility function converts performance on that attribute (a cost, for instance) into a utility that reflects the decision maker's preferences about that attribute. Moreover, the multiattribute utility function that combines the utility functions reflects the decision maker's preferences about the relative importance of the different attributes.

For example, a MAU model was used to help the U.S. Department of Energy (DOE) and Russian scientists evaluate alternatives for the disposition of surplus weapons-grade plutonium (Dyer et al. 1998; Butler et al. 2005). Keeping this plutonium secure is important to protect the public and the environment from the harmful effects of radiation and to prevent terrorists from acquiring it. Thirteen different disposal options were considered on a large set of objectives. The nonproliferation objectives included minimizing the opportunities for others to steal the plutonium, minimizing the ability of anyone to divert the plutonium during processing, maximizing the difficulty of recovering the plutonium after processing, fostering international cooperation, and minimizing the time required to start and end the disposition. The environment, safety, and health objectives included minimizing the impact on the public and workers, minimizing the impact on the environment, and minimizing the impact on the economy. There were also objectives to minimize investment and life-cycle costs. The analysts and DOE experts created a MAU model by assessing the single-attribute utility functions and the weights associated with each objective. They also created a decision tree to consider how the Russians could respond to the U.S. decision, which led to the DOE adopting a hybrid approach that pursued two alternatives in parallel. Later, the DOE decided to convert the plutonium into mixed oxide fuel. The U.S. team then worked with Russian scientists to build a similar MAU model that reflected the Russian preferences and used this to evaluate 12 alternatives. This analysis led the Russians to select the same type of disposition. As Butler et al. (2005) noted, using the MAU model allowed to discussion of the alternatives to consider multiple points of view (without emotion) and maintain a balanced perspective.

1.5.2 Game Theory

Game theory studies multiple decision makers who decide independently, but the outcome is determined by their joint decision. In some cases, whatever one player gains, the other player loses (a zero-sum game). In mixed-motive games, the payoffs to each player are more general. Nash (1951), a classic reference in game theory, studied non-cooperative games. Other classic texts are von Neumann and Morgenstern (2007) and Luce and Raiffa (1957). Raiffa et al. (2007) placed game theory in the context of negotiations, and the collection by Bier and Azaiez (2009) presented applications to analyzing security threats.

When we consider games, we often think of activities such as chess or tic-tac-toe, where the players alternate their moves. Unless the state space gets too large, as it does in chess, such games are relatively easy to analyze because, when a player needs to decide which move to make, they know everything.

In simultaneous games like rock-paper-scissors, neither player is sure what the other one will do. In the traditional case, each player knows what the other could do, and both players know the payoff matrix, which describes the reward (or penalty) that each player will receive (pay) given their joint decisions.

In the domain of homeland security, game theory is a valuable tool because, as Bier (2006) concluded, "protecting against intentional attacks is fundamentally different from protecting against accidents or acts of nature." The terrorists' ability to evaluate defensive measures and then choose the best way to attack requires considering game theoretic approaches.

In some situations, the agents (the attackers and the defenders) move simultaneously, but in others they move sequentially.

Game theory is especially useful (and relatively easy to use) for making decisions in the following situations (Cox 2009): (1) the defender allocates his resources (e.g., money, personnel, and equipment) to defend targets; (2) the attacker, after observing the defender's allocation of defensive resources, allocates his resources to attack targets; and (3) each player receives a consequence. The defender's consequences may include the number of people killed or injured, the property destroyed, and the psychological harm and lifestyle disruption.

Hausken (2011) not only emphasized that different types of models are needed for these different situations but also demonstrated that incorrectly assuming that the agents move simultaneously when, in reality, one agent moves first can lead one to choose a poor alternative with disastrous consequences.

An example of a simultaneous game was presented by Gaver et al. (2009), who considered a counterterrorism agent searching for a terrorist in a crowd of neutral individuals. The counterterrorism agent must decide how much time to spend investigating each individual encountered. Spending too little time results in many mistakes, which waste time; spending too much time reduces the rate at which individuals are investigated. Both cases increase the total time needed to intercept and neutralize the terrorist. Meanwhile, the terrorist is looking for a target whose value is greater than a threshold that he must select. A game-theoretic model was used to identify the optimal investigation time and to show that the optimal time is robust to uncertainties in certain parameters of the model.

1.6 What Is the Best Solution? Solving Problems with Optimization

Optimization is a key technique in operations research. Optimization is essentially a search technique that is most appropriate when the set of possible solutions is large and complex and evaluating a possible solution requires significant effort (Bonabeau 2003). An optimization problem is usually defined by a set of decision variables, a set of constraints, and an objective function. The decision variables are those that need to be determined; selecting values for the decision variables specifies a particular solution. The constraints are relationships (expressed as mathematical equations) between the decision variables that determine whether a possible solution is feasible. A possible solution is feasible if and only if each and every constraint is satisfied. The objective function, also a mathematical expression, calculates the performance measure that is to be optimized.

It is not always necessary or possible to write the optimization problem using mathematical expressions. In some types of optimization problems, it is difficult or tedious to write down the constraints explicitly. For instance, in the vehicle routing problem, a feasible solution cannot have partial routes (called "subtours") that do not begin and end at the depot. A very large set of constraints is needed to express this restriction. In simulation optimization problems, simulation models are required to evaluate the objective function or the constraints.

Using optimization to solve a problem or make a decision begins with identifying the decision variables, formulating the constraints, and determining the objective function. The next step is to select a technique (the "solver") for finding the optimal solution. It is desirable to select a solver that can exploit any special structure in the formulation. For instance, if all of the decision variables are continuous real variables and the constraints and objective function are all linear expressions, then one can use specialized solvers for linear programming.

The existence of special-purpose solvers often influences the formulation. For instance, approximating a complicated objective function by a simpler linear one allows one to solve the problem more quickly. When the complexity of the problem makes finding the optimal solutions a difficult and time-consuming process, heuristic searches such as simulated annealing and genetic algorithms can be used to look for high-quality solutions; these will require less time but the best solution found may not be optimal.

After selecting and running the solver to find the optimal solution, it is often useful to conduct a sensitivity analysis to determine how any changes to the input parameters would affect the values of the optimal solution and optimal objective function value.

The work described in the following sections on mathematical programming, facility location problems, and vehicle routing problems all fall into the general area of optimization.

Preventing terrorists from exploding bombs onboard commercial flights requires screening the baggage that the passengers check and carry on these planes. Because the number of baggage screening devices is limited, not all bags can be appropriately screened, however. This leads to two different measures of the risk of a bombing (1) the number of flights with unscreened baggage onboard and (2) the number of passengers on these flights (Jacobson et al. 2005). Given a set of bags that need to be screened (but not enough time and resources to screen all of them), one needs to select a subset that can be screened in time. Jacobson et al. (2003, 2005) presented optimization models for designing screening systems that minimize these two risk measures. Their results on real-world cases indicated that minimizing the number of flights with unscreened baggage is not difficult and also reduces the number of passengers on these flights and so should be considered a superior risk minimization strategy.

Similarly, Candalino et al. (2004) used optimization to find good strategies for airport security operations. The decision variables were the configuration of baggage screening security devices: a specific type of security screening device for each level of baggage screening. The objective function calculated the costs of purchasing and operating baggage screening security devices and the costs of screening errors (including false alarms). The goal was to minimize the total cost. The constraints were expressions that calculate the screening error rates and the number of devices (of the selected type) at each level. Also, because there are 40 different types of security screening devices, the value of each decision variable was an integer in the range from 1 to 40.

Although this problem could have been formulated explicitly using mathematical expressions, there is no solver that can find an optimal solution in reasonable time. Thus, Candolino et al. used simulated annealing to find high-quality solutions. In general, a simulated annealing procedure begins with one solution and randomly selects another solution that is "near" the current solution. If the new solution is better, then it is accepted. If not, the procedure accepts the new solution with a probability that depends upon how bad it is (a worse solution has a lower probability of being accepted). This continues until the search converges. Candolino et al. showed that the best screening strategy depended upon the threat level. When the threat level is high, the overall cost was less when more precise (but more expensive) screening devices are purchased, as this lowers the operational costs and the costs of screening errors.

1.6.1 Mathematical Programming

Mathematical programming is a class of optimization techniques that relies upon formally expressing the optimization problem as a set of mathematical expressions that represent the objective function and the constraints of the problem. The most important classes of mathematical programming include linear programming, network models, integer programming, dynamic programming, and nonlinear programming (Bradley et al. 1977).

Because mathematical programming is a type of optimization approach, the general technique follows that used for optimization. As mentioned earlier, for some types of mathematical programming problems, specialized solvers exist to find optimal solutions quickly. The simplex method is a well-known technique for solving linear programs. When specialized solvers are not available, one can choose from a variety of general purpose optimization algorithms and search procedures.

A wide variety of mathematical programming techniques have been applied to problems related to homeland security. The following paragraphs will provide some examples.

Lim et al. (2009) used linear programming to formulate the problem of quickly evacuating a metropolitan area in advance of a hurricane. The decision variables were the number of people who move between pairs of adjacent locations in a time period. The objective function was to maximize the number of people who reach a safe place during the evacuation period (before the hurricane arrives). The constraints ensured the conservation of flow. A given evacuation period was feasible if a solution exists in which everyone reaches a safe place.

Cormican et al. (1998) considered the problem of network interdiction, in which an adversary (e.g., drug smugglers) will use the arcs of the network (e.g., roads) to move material. The decision maker can interdict arcs, which prevents the adversary from using them (e.g., by installing roadblocks). The decision maker wants to select a set of arcs that, when interdicted, minimizes the maximum flow through the network. The decision variable is the set of arcs to interdict. Interdicting an arc requires resources (e.g., money), which are limited. This limitation imposes the following constraint: the total resources spent to interdict arcs must be less than or equal to the resources available (e.g., the budget). Because the problem is difficult to solve optimally, Cormican et al. developed a specialized algorithm to find high-quality solutions.

In a binary integer program, the decision variables are all "yes/no" variables that indicate whether an object is selected for some reason. Nehme and Morton (2009) formulated a binary integer program for deciding which border checkpoints should have nuclear detectors. The objective function was to minimize the probability that a smuggler avoids detection.

Lim and Smith (2007) studied a version of the network interdiction problem in which there were multiple commodities and solved it by reformulating the problem as a mixed integer program. That is, there were both continuous and binary variables. They solved the problem using a commercial optimization solver.

Bansal and Kianfar (2010) formulated a nonlinear programming problem for deciding where a surveillance camera should point when different areas are likely to generate interesting objects. The problem had two decision variables (the two-dimensional coordinates that describe where the surveillance camera should point). The objective function was to maximize the "value" of the area covered by the camera. Bansal and Kianfar presented a branch-and-bound algorithm that implicitly searched the set of feasible solutions.

Martonosi (2011) studied the problem of switching servers between different queues. For example, managers may move security personnel from one passenger screening area to another where the number of passengers waiting is very long. Unfortunately, because the passenger arrival rates can change during the day and time is wasted when switching servers, reducing the average waiting time is difficult, and simple rules are not effective. Martonosi presented an approximate dynamic programming approach that can find, for a two-queue system, the optimal policy. This policy describes, given the queue lengths at a point in time, how many (if any) servers should be switched from one queue to the other.

1.6.2 Facility Location Problems

Facility location problems are a class of optimization problems concerned with finding the best locations to place facilities. In the context of homeland security, the facilities to be placed include emergency warning sirens (cf. Current and O'Kelly 1992); prepositioned caches of medical supplies that will be delivered soon after an emergency; Receipt, Storage, and Stage (RSS) facilities that will handle the supplies being delivered to the site of an emergency; PODs where residents will pick up medication; and disaster recovery centers where officials provide recovery assistance to victims.

In general, facilities should be located near those who need their services (the demand points). If the number of facilities is fixed, then the objective is to minimize the distance from the demand points. The n-median problem minimizes the total distance between demand points and facilities. The n-center problem minimizes the maximum distance between demand points and facilities. In both types of problems, the demand points may have weights that correspond to the amount of demand or the cost of satisfying that demand.

Otherwise, the objective is to minimize the number of facilities needed while satisfying constraints on the distance from demand points to facilities. This is known as the covering model (or problem). The two primary subclasses are the Location Set Covering Problem (LSCP) and the Maximal Covering Location Problem (MCLP).

Dekle et al. (2005) studied the problem of minimizing the total number of disaster recovery centers (temporary offices where officials provide recovery assistance to victims) in Alachua County, Florida, subject to each county resident being close to a facility (that is, the distance between a demand point and the closest facility must be less than a given threshold). The problem, a covering model, was solved for three possible choices of the distance threshold: 10, 15, and 20 miles to identify optimal locations. Then, buildings near these coordinates were evaluated as potential disaster recovery center sites. Because the underlying road network resembles a grid, the distance between demand points and facilities was estimated using a rectilinear distance measure. In addition, the number of demand points and possible facility sites was reduced by aggregating those that were very near each other.

Jia et al. (2007) presented a general facility location problem for a Large-scale Emergency Medical Service (LEMS). For a particular scenario, the problem is to locate n facilities so that the required number of facilities services each demand point with the same quality. The problem generalizes the covering, n-median, and n-center problems. The covering model finds a solution that maximizes the demand that can be covered. The n-median problem minimizes the sum of the service distances for all demand points; the n-center problem minimizes the maximum service distance.

Jia et al. illustrated the use of these models by formulating location problems for dirty bomb, anthrax, and smallpox terrorist attacks. For the dirty bomb attack, supplies such as protective equipment and anti-radioactive drugs must be prepositioned at facilities in the region. For the anthrax attack, RSS facilities must be placed to minimize the maximum distance from demand points. For the smallpox attack, the locations of local caches of vaccine for first responders is determined by using a n-center model, and the location of RSS facilities is determined by using a n-median model.

1.6.3 Vehicle Routing Problems

Vehicle routing problems are a class of optimization problems concerned with finding the best way to route a set of vehicles to deliver material to a number of customers (Toth and Vigo 2002). Solving such problems is especially important when it is necessary to deliver emergency supplies and medication to the victims of a terrorist attack quickly. Most of the research on vehicle routing problems has considered the objective of minimizing the number of vehicles needed, the total travel time, or the total transportation cost. In the context of homeland security, however, the objective is to meet demand quickly given the limited resources available. Delays in meeting demand cause additional suffering and casualties. Vehicle routing problems also occur in the context of delivering disaster relief supplies.

For example, Shen et al. (2009b) considered the problem of routing vehicles to meet demand in an emergency. In this problem, each demand site has a deadline, and it may not be possible to visit all sites in time. The objective, therefore, is to minimize the unmet demand. The travel times between sites and demand at each site are unknown. Because probability distributions for the travel times and demand are given, however, the formulation includes constraints that require a solution to be feasible (with respect to deadlines and demand) most of the time (technically, the model is a chance-constrained model). Shen et al. (2009a) included this problem as the first stage of a two-stage approach. Solving the first stage before the emergency occurs generates preplanned routes, which can be useful for exercises and training. The second stage is solved after the emergency occurs; at this point, the actual travel times and demands are known, and the original solution can be adjusted by modifying the delivery quantities, skipping low-demand sites, or finding a completely new solution.

The inventory slack routing problem (ISRP) is a much different problem that occurs because not all material is available initially and demand sites use material over time (it is not all required at a single point in time) (Herrmann et al. 2009; Montjoy and Herrmann 2010). Vehicles must return to the central depot to get more material when it arrives and deliver it to the demand sites. It is critical to deliver sufficient material in a timely manner so that sites can continue operating without interruption. As a hedge (or buffer) against the uncertainty in the travel times, it is desirable to have slack in the deliveries (that is, the deliveries occur earlier than needed). In order to be fair to all demand sites, the objective of the ISRP is to maximize the minimum slack. Herrmann et al. (2009) separated the problem into a routing subproblem and a scheduling subproblem and solved each subproblem using fast heuristics; Montjoy and Herrmann (2010) adopted the same separation but solved the routing subproblem using a search heuristic.

1.7 Summary

This chapter described a diverse set of OR techniques that can be used to help homeland security professionals make better decisions. Unlike previous reviews, which focused on the range of applications for OR techniques, this review briefly described relevant OR techniques and presented examples of how they have been used to model homeland security problems.

Homeland security, like other domains such as manufacturing and transportation, involves decisions related to planning, monitoring, and controlling operations at many levels. OR techniques can help decision makers understand what has happened by analyzing large data sets and monitoring time series data; consider what could happen by analyzing risk, modeling queueing systems, and simulating complex systems; improve decision making by analyzing decisions and considering what others might do; and find better solutions by formulating and solving optimization problems.

Some problems in homeland security require only the straightforward application of common OR techniques and can be done by anyone who can use a spreadsheet. Other problems require significant extensions to existing techniques or specialized skills to formulate and solve the problem.

Homeland security professionals can use this chapter to understand what operations research can offer and identify specific techniques that may be relevant to their problems. The chapter includes references to essential books and articles that can provide more information about the techniques. The examples (with citations to references for more information) illustrate how specific techniques have been applied to homeland security problems.

References

Aaby K, Herrmann JW, Jordan C, Treadwell M, Wood K (2006a) Montgomery County's Public Health Service uses operations research to plan emergency mass-dispensing and vaccination clinics. Interfaces 36(6):569–579

Aaby K, Abbey R, Herrmann JW, Treadwell M, Jordan C, Wood K (2006b) Embracing computer modeling to address pandemic influenza in the 21st century. J Public Health Manag Pract 12(4):365–372

Aleman DM, Wibisono TG, Schwartz B (2011) A nonhomogeneous agent-based simulation approach to modeling the spread of disease in a pandemic outbreak. Interfaces 41(3):301–315

Altay N, Green WG (2006) OR/MS research in disaster operations management. Eur J Oper Res 175:475–493

Bakuli DL, Smith JM (1996) Resource allocation in state-dependent emergency evacuation networks. Eur J Oper Res 89(3):543–555

Bansal M, Kianfar K (2010) An exact algorithm for coverage problem with a single rectangle. In: Johnson A, Miller J (eds) Proceedings of the 2010 Industrial Engineering Research Conference, Cancun, 5–9 June 2010

Bier V (2006) Game-theoretic and reliability methods in counterterrorism and security. In: Wilson AG, Wilson GD, Olwell DH (eds) Statistical methods in counterterrorism. Springer, New York

Bier VM, Azaiez MN (eds) (2009) Game theoretic risk analysis of security threats. Springer, New York

Bonabeau E (2003) Don't trust your gut. Harv Bus Rev 81(5):116–123

Box GEP, Jenkins GM, Reinsel GC (1994) Time series analysis, 3rd edn. Prentice Hall, Englewood Cliffs, NJ

Bradley SP, Hax AC, Magnanti TL (1977) Applied mathematical programming. Addison-Wesley, Reading, MA

Brown GG, Cox LA Jr (2011) How probabilistic risk assessment can mislead terrorism risk analysts. Risk Anal 31(2):196–204

Butler JC, Chebeskov AN, Dyer JS, Edmunds TA, Jia J, Oussanov VI (2005) The United States and Russia evaluate plutonium disposition options with multiattribute utility theory. Interfaces 35(1):88–101

Candalino TJ, Jacobson SH, Kobza JE (2004) Designing optimal aviation baggage screening strategies using simulated annealing. Comput Oper Res 31(10):1753–1767

Chen W, Hsu S, Shen H (2005) Application of SVM and ANN for intrusion detection. Comput Oper Res 32(10):2617–2634

Clemen RT, Reilly T (2001) Making hard decisions with decision tools. Duxbury, Pacific Grove, CA

Cormican KJ, Morton DP, Wood RK (1998) Stochastic network interdiction. Oper Res 46:184–197

Cox LA Jr (2009) Game theory and risk analysis. Risk Anal 29(8):1062–1068

Current J, O'Kelly M (1992) Locating emergency warning sirens. Decis Sci 23(1):221–234

Das TK, Savachkin AA, Zhu Y (2008) A large-scale simulation model of pandemic influenza outbreaks for development of dynamic mitigation strategies. IIE Trans 40:893–905

Defense Science Board Advisory Group on Defense Intelligence (2009) Operations research applications for intelligence, surveillance, and reconnaissance (ISR). Office of the Under Secretary of Defense for Acquisition, Technology, and Logistics, Washington, DC

Dekle J, Lavieri MS, Martin E, Emir-Farinas H, Francis RL (2005) A Florida county locates disaster recovery centers. Interfaces 35(2):133–139

Denizel M, Usdiken B, Tuncalp D (2003) Drift or shift? Continuity, change, and international variation in knowledge production in OR/MS. Oper Res 51(5):711–720

Department of Homeland Security (DHS) (2011) Programs & projects. http://www.dhs.gov/files/scitech.shtm. Accessed 1 June 2011

Dyer JS, Edmunds T, Butler JC, Jia J (1998) A multiattribute utility analysis of alternatives for the disposition of surplus weapons-grade plutonium. Oper Res 46(6):749–762
Ezell BC, Bennett SP, Von Winterfeldt D, Sokolowski J, Collins AJ (2010) Probabilistic risk analysis and terrorism risk. Risk Anal 30(4):575–589
Fienberg SE (2008) Homeland insecurity: data mining, privacy, disclosure limitation, and the hunt for terrorists. In: Chen H, Reid E, Sinai J, Silke A, Ganor B (eds) Terrorism informatics. Springer, New York
Gaver DP, Glazebrook KD, Jacobs PA (2009) Search for a malevolent needle in a benign haystack. In: Bier VM, Azaiez MN (eds) Game theoretic risk analysis of security threats. Springer, New York
Gilliam RR (1979) An application of queueing theory to airport passenger security screening. Interfaces 9(4):117–122
Hall RW (1991) Queuing methods for services and manufacturing. Prentice Hall, Englewood Cliffs, NJ
Hastie T, Tibshirani R, Friedman J (2009) The elements of statistical learning: data mining, inference, and prediction, 2nd edn. Springer, New York
Hausken K (2011) Strategic defense and attack of series systems when agents move sequentially. IIE Trans 43(7):483–504
Herrmann JW (2008) Disseminating emergency preparedness planning models as automatically generated custom spreadsheets. Interfaces 38(4):263–270
Herrmann JW, Lu S, Schalliol K (2009) A routing and scheduling approach for planning medication distribution. In: Proceedings of the 2009 Industrial Engineering Research Conference, Miami, FL, 30 May–June 3 June 2009
Institute for Operations Research and the Management Sciences (2011) About operations research. http://informs.org/About-INFORMS/About-Operations-Research. Accessed 1 June 2011
Jacobson SH, Kobza JE, Nakayama MK (2000) Sampling procedure to estimate risk probabilities in access-control security systems. Eur J Oper Res 122(1):123–132
Jacobson SH, Virta JL, Bowman JM, Kobza JE, Nestor JJ (2003) Modeling aviation baggage screening security systems: a case study. IIE Trans 35(3):259–269
Jacobson SH, McLay LA, Kobza JE, Bowman JM (2005) Modeling and analyzing multiple station baggage screening security system performance. Naval Res Logist 52(1):30–45
Jain S, McLean CR, Lee YT (2007) Towards standards for integrated gaming and simulation for incident management. In: Proceedings of the 2007 summer computer simulation conference (SCSC). Society for Computer Simulation International, San Diego, CA, pp 1213–1222
Jia H, Fernando O, Maged D (2007) A modeling framework for facility location of medical services for large-scale emergencies. IIE Trans 39(1):41–55
Jonas J, Harper J (2006) Effective counterterrorism and the limited role of predictive data mining. Policy Anal 584:1–12. http://www.cato.org/pubs/pas/pa584.pdf. Accessed 9 Feb 2011
Kaplan EH (2010) Terror queues. Oper Res 58(4):773–784
Kobza JE, Jacobson SH (1996) Addressing the dependency problem in access security system architecture design. Risk Anal 16(6):801–812
Kobza JE, Jacobson SH (1997) Probability models for access security system architectures. J Oper Res Soc 48(3):255–263
Krings A, Azadmanesh WA (2005) A graph based model for survivability applications. Eur J Oper Res 164(3):680–689
Law AM (2007) Simulation modeling & analysis, 4th edn. McGraw-Hill, Boston
Li H, Apostolakis GE, Gifun J, VanSchalkwyk W, Leite S, Barber D (2009) Ranking the risks from multiple hazards in a small community. Risk Anal 29(3):438–456
Lim C, Smith JC (2007) Algorithms for discrete and continuous multicommodity flow network interdiction problems. IIE Trans 39(1):15–26
Lim GJ, Zangeneh S, Baharnemati MR, Assavapokee T (2009) A sequential solution approach for short notice evacuation scheduling and routing. In: Proceedings of the 2009 industrial engineering research conference, Miami, FL, 30 May–3 June 2009
Lombardo JS, Burkom H, Pavlin J (2004) ESSENCE II and the Framework for evaluating syndromic surveillance systems. MMWR 53(Suppl):159–165. http://www.cdc.gov/mmwr/preview/mmwrhtml/su5301a30.htm. Accessed 31 May 2011

Luce RD, Raiffa H (1957) Games and decisions. Wiley, New York

Martonosi S (2011) Dynamic server allocation at parallel queues. IIE Trans 43(12):863–877

Martonosi SE, Barnett A (2006) How effective is security screening of airline passengers. Interfaces 36(6):545–552

McCormick PM, McNeill G, Hendrix D, McCormick T (2010) Complexity, terror and Murphy's law. OR/MS Today 37(6):26–33

Meredith JR (2001) Reconsidering the philosophical basis of OR/MS. Oper Res 49(3):325–333

Michalak S, Wilson G (2006) Modeling and parameterization for a smallpox simulation study. In: Wilson AG, Wilson GD, Olwell DH (eds) Statistical methods in counterterrorism. Springer, New York

Miller G, Randolph S, Patterson JE (2006) Responding to bioterrorist smallpox in San Antonio. Interfaces 36(6):580–590

Modarres M (2006) Risk analysis in engineering. CRC, Boca Raton, FL

Montjoy A, Herrmann JW (2010) Adaptive large neighborhood search for the inventory slack routing problem. In: Proceedings of the 2010 Industrial Engineering Research Conference, Cancun, 5–9 June 2010

Moore M, Wermuth MA, Castaneda LW, Chandra A, Noricks D, Resnick AC, Chu C, Burks JJ (2010) Bridging the gap: developing a tool to support local civilian and military disaster preparedness. Rand Corporation, Center for Military Health Policy Research, Santa Monica, CA

Nash J (1951) Non-cooperative games. Ann Math 54(2):286–295

Nehme MV, Morton DP (2009) Tightening a network interdiction model. In: Proceedings of the 2009 Industrial Engineering Research Conference, Miami, FL, 30 May–3 June 2009

Newell GF (1982) Applications of queueing theory, 2nd edn. Chapman and Hall, London

O'Hara S, McLean CR, Lee YT (2010) Modeling and simulation for emergency management and health care systems: workshop summary (NISTIR 7684). National Institute of Standards and Technology, Gaithersburg, MD

Paté-Cornell ME (2009) Games and risk analysis. In: Bier VM, Azaiez MN (eds) Game theoretic risk analysis of security threats. Springer, New York

Paté-Cornell ME, Guikema SD (2002) Probabilistic modeling of terrorist threats: a systems analysis approach to setting priorities among countermeasures. Military Oper Res 7(4):5–23

Qin J, Zhou Y, Reid E, Chen H (2008) Studying global extremist organizations' internet presence using the dark web attribute system: a three region comparison study. In: Chen H, Reid E, Sinai J, Silke A, Ganor B (eds) Terrorism informatics. Springer, New York

Raiffa H, John R, David M (2007) Negotiation analysis: the science and art of collaborative decision making. Belknap Press of Harvard University Press, Cambridge, MA

Shahar Y (2008) Toward a target-specific method of threat assessment. In: Chen H, Reid E, Sinai J, Silke A, Ganor B (eds) Terrorism informatics. Springer, New York

Shen Z, Dessouky MM, Ordonez F (2009a) A two-stage vehicle routing model for large-scale bioterrrorism emergencies. Networks 54:255–269

Shen Z, Ordóñez F, Dessouky MM (2009b) The stochastic vehicle routing problem for minimum unmet demand, Optimization and logistics challenges in the enterprise. Springer series on optimization and its applications. Springer, Heidelberg

Shmueli G, Fienberg SE (2006) Current and potential statistical methods for monitoring multiple data streams for biosurveillance. In: Wilson AG, Wilson GD, Olwell DH (eds) Statistical methods in counterterrorism. Springer, New York

Sparks R, Carter C, Graham P, Muscatello D, Churches T, Kaldor J, Turner R, Zheng W, Ryan L (2010) Understanding sources of variation in syndromic surveillance for early warning of natural or intentional disease outbreaks. IIE Trans 42(9):613–631

Stoto MA, Fricker RD Jr, Jain A, Diamond A, Davies-Cole JO, Glymph C, Kidane G, Lum G, Jones LV, Dehan K, Yuan C (2006) Evaluating statistical methods for syndromic surveillance. In: Wilson AG, Wilson GD, Olwell DH (eds) Statistical methods in counterterrorism. Springer, New York

Stroud P, Del Valle S, Sydoriak S, Riese J, Mniszewski S (2007) Spatial dynamics of pandemic influenza in a massive artificial society. Journal of Artificial Societies and Social Simulation 10(4)9

Toth P, Vigo D (eds) (2002) The vehicle routing problem, SIAM monographs on discrete mathematics and applications. SIAM Publishing, Philadelphia

von Neumann J, Morgenstern O (2007) Theory of games and economic behavior, Commemorative edn. Princeton University Press, Princeton, NJ

Wright PD, Matthew JL, Robert LN (2006) A survey of operations research models and applications in homeland security. Interfaces 36(6):514–529

Chapter 2
Operations Research and Homeland Security: Overview and Case Study of Pandemic Influenza

Richard C. Larson, Anna Teytelman, and Stan Finkelstein

Abstract This chapter starts with a brief review of the birth of operations research (OR) in a war-focused homeland security setting. In more frequent non-war settings, homeland security requires diligent planning for and responding to low probability, high consequence (LPHC) events. Resulting decisions are described by time frame, from long-term planning decisions to minute-to-minute operational decisions. After a brief review of OR tools and techniques and of related literature, the chapter then provides details of recent OR research by the authors into a major threat to homeland security: pandemic influenza. In the context of the 2009 H1N1 pandemic flu, we examine a supply chain problem in pandemic flu response: the manufacture, delivery, and administering of flu vaccine. Using a combination of axiomatically derived OR models and data obtained from a representative sample of states of the USA, we evaluate the effectiveness of the deployment of flu vaccines in 2009. We show that, for many states, the vaccines arrived far too late to be useful. We suggest alternative vaccine allocation policies that could dramatically increase the numbers of flu infections averted. We offer practical takeaways for those whose responsibility it is to design homeland security response strategies for their own states and communities.

2.1 Introduction and Overview

Operations research (OR) is the world's most important invisible profession! OR is everywhere and yet not seen. It is in logistics, health care, urban systems, retailing, energy, mining, entertainment, and much more. OR is concerned with conceptualizing and implementing mathematical models to help solve planning and operating problems arising in the public and private sectors. These models are

R.C. Larson (✉) • A. Teytelman • S. Finkelstein
Massachusetts Institute of Technology, Cambridge, MA 02139, USA
e-mail: rclarson@mit.edu

J.W. Herrmann (ed.), *Handbook of Operations Research for Homeland Security*, International Series in Operations Research & Management Science 183,
DOI 10.1007/978-1-4614-5278-2_2,

used to provide decision makers with better insights into their problems and to assist them in selecting the most effective courses of action. In a nutshell, OR uses science and mathematics to provide insight to decision makers, hopefully leading to better decisions. Here, *a decision is an allocation of resources.*

Homeland security gave birth to OR as we know it today, and thus the OR-Homeland-Security connection is a most appropriate focus for this chapter. OR was born in Great Britain and in the USA at the beginning of World War II, by physicists P.M.S. Blackett in the UK and Philip M. Morse in the USA. The resources to be allocated using OR methods were men and material associated with the war effort (Morse and Kimball 2003). There were numerous successes, including the optimal deployment of the then new, scarce and expensive radar stations in England, to optimal search strategies for destroyers in the North Atlantic—seeking out enemy submarines. B.O. Koopman's work on search theory was so important that it was classified Top Secret and published in the open literature only in the late 1950s, fully 15 years after its creation (Koopman 1957).

In this chapter we first review briefly some aspects of OR as applied to current definitions of homeland security. In our view, homeland security requires diligent planning for and responding to low probability, high consequence (LPHC) events—such as severe acts of Mother Nature, damaging human-caused accidents and terrorist attacks. Much has been written about OR as applied to homeland security, and we provide some guidance to the literature and some illustrative takeaways. Second, we go into detail applying OR to the planning for and responding to a major LPHC event—pandemic influenza. This work has been carried out over the last 5 years by the authors and our associates at MIT and at the Harvard School of Public Health.

The initial motivation was the threat that the highly lethal H5N1 flu virus, often called "bird flu," would mutate and become highly transmittable from human to human. Leaders in the international health community feared a repeat of the infamous 1918 "Spanish Flu" pandemic, which killed about 50,000,000 people worldwide in a matter of months. The arrival of the H1N1 pandemic virus in 2009 provided a "dress rehearsal" for such a future killer contagion. The matter is even more urgent now, as the H5N1 flu could, it is feared, be used as a bio-terrorism weapon of mass destruction. This possibility has been brought more sharply into focus now that scientists have discovered sequences of mutations of H5N1 that make it highly transmittable and yet retain its high lethality (Greenfieldboyce 2011).

2.2 OR-Informed Decisions for Homeland Security

We focus on LPHC events and decisions to be made. Decisions come in three flavors: strategic, tactical, and operational. The words are taken from the military roots of OR. They basically relate to the time frames involved: long-range (strategic), medium range (tactical), and real time (operational). But the issues are generic to all types of decision situations. The coach of a football team shares the same three decision flavors:

- *Strategic*: How should I build my team by player drafts and trades to have the most potent team possible, subject to all sorts of constraints such as budgets?
- *Tactical*: On a week-to-week basis, which players should I field on game day and which ones should I place off-roster?
- *Operational*: It is fourth down and one yard to go on our opponent's 48-yard line. Do I punt the ball to the other team or try to make a first down, recognizing that if we fail, we give the opposing team terrific field position?

OR professionals have a rich set of tools in their toolbox to deal with these types of decisions for football coaches such as Bill Belichick of the New England Patriots (Leonhardt 2004) and for the professionals who have to plan for and respond to LPHC events.

Here is one possible set of decisions related to planning for and responding to LPHC events:

- *Strategic*: How many resources should I have that can be devoted to a given type of LPHC event in my state? Think hurricane, earthquake, blizzard, flood, major industrial accident, or terrorist attack. And what additional resources can I call on if my own resources become overwhelmed?
- *Tactical*: How frequently should I plan for disaster drills, using not only my own resources but neighboring resources under a mutual aid agreement?
- *Operational*: A magnitude 7.4 earthquake centered a few miles south of San Francisco has just devastated parts of the city. Right now, where should I place my limited rescue resources, how do I triage for injuries, and what do I communicate to the public?

Not every aspect of these decisions is guided by OR modeling and analysis. But a significant fraction of them can be. Others are guided by experience, craft knowledge, and common sense.

As one illustrative application of problem framing OR tools, let us examine queueing theory. A queue is a waiting line, and in a disaster, queues will exist in many places. A queue appears whenever arriving "customers" require more service than there is capacity to serve those customers. In supermarkets and coffee shops, queues may be annoying and delay you for a few minutes of your life, but in LPHC disasters they may be long and life threatening. Almost invariably, an LPHC disaster will result in queues having far more service requirements than there is locally available capacity to serve them. Managing these queues effectively is vitally important and requires intelligent triage to prioritize the queue, separate people requiring emergency service into different priority levels, and treat them accordingly. Sometimes the decisions are difficult and tragic, as they were in Pearl Harbor on December 7, 1941, when—to ease pain—nurses provided morphine to critically injured servicemen, whose foreheads were then marked by the nurse's lipstick with an "M," "C," or "D," representing "morphine," "critical," or "deceased," respectively. Improvisation is often required. Rapid assessment of the situation is vital to selecting and implementing an intelligent triage policy and then administering the best possible service to the sick and wounded. Queueing theory, studied and internalized by planners ahead of time, can lay bare the essentials of service capacity, anticipated delays in queue as a function of the number of "customers," the consequences of alternative prioritization or triage strategies, the numbers of required servers, and more. The Walt Disney World theme parks employ 15–20 OR analysts to study and "optimize" queues in their theme parks. Planners for LPHC events must do the same—for situations much more critical than amusement park ride delays.

Queue delays grow highly nonlinearly with the numbers of "customers" and in fact can exponentially explode, resulting in intolerably huge queues. Even queues operating in non-emergency situations can become large due to the variability in the system, i.e., unscheduled times of arrivals of customers and widely varying service requirements—meaning service times. Anyone planning for disasters should have a staff member study the essentials of queueing theory to help create a feasible and effective response plan that treats the most critical "customers" first. The plan must also include the procedures to bring in additional queue "servers," often via local and regional mutual aid agreements. As teaching guides, accessible applications of queueing theory to day-to-day public safety systems are readily available (Larson 1972).

Queuing theory is but one tool in the OR tool box. There are many others, including optimization methods such as linear programming and dynamic programming, transportation network analysis, and decision theory. Most of these tools have relevance to planning for and responding to LPHC events.

For those who wish to explore further the general application of OR to LPHC events, we suggest some publications previously written by one or more of us (usually with other coauthors). And each paper contains many references that are also important for LPHC events. The first is "Disasters: Lessons from the Past 105 Years" (Eshghi and Larson 2008). This overview paper shows an alarming reported increase in frequency of disasters over the most recent century-plus time period, some of the increase due to improved monitoring technologies (e.g., for earthquakes) and some due to population growth where people now live in more-disaster-prone places. The second is "Responding to Emergencies: Lessons Learned and the Need for Analysis" (Larson et al. 2006). This paper contains historical reviews of five major emergencies—the Oklahoma City bombing (1995), the crash of United Airlines Flight 232 (1989), the sarin attack in the Tokyo subway (1995), Hurricane Floyd (1999), and Hurricane Charlie (2004). The paper draws lessons from these five LPHC events and outlines required additional OR research that is needed to cope better with similar disasters in the future. The third is "Decision Models for Emergency Response Planning" (Larson 2005). This book chapter covers OR not only as it applies to response to major disasters including terrorism but also to routine public safety operations (police, fire, and ambulance) and to dangerous operations such as transporting hazardous materials.

2.3 Pandemic Flu: Background

Those who have heard of OR but have not studied it may think that the only applications of OR to pandemic flu are strictly operational. A popular idea for instance is to apply queueing theory to the system of vaccine dispensing in a public clinic, with the goal of dispensing the vaccine quickly and fairly. While this certainly can be done, and should be done, it represents only a tiny fraction of the myriad decision problems that arise when planning for and responding to pandemic influenza. And many, perhaps most, of these decision problems can be better informed by OR problem framing and analysis.

In considering vaccines to immunize people against a particular flu strain, we expand our OR problem framing beyond the public clinic offering the vaccine all the way back to the origins of the new flu virus, the creation of a new vaccine, and its distribution and administration throughout the country. This is a process that involves designing a new product, manufacturing it in sufficient quantities to satisfy customer demand, transporting it to regional distribution centers, deploying it to local centers (such as flu clinics), and then delivering it to customers (susceptible members of the population). This is a typical OR supply chain analysis problem. It requires analysis of the *entire system*, not just the queueing at the end of the process. Clinic queueing may be optimized, almost reduced to zero in the clinics, but that is of no consequence to citizens who are susceptible to the flu if the vaccine does not arrive in time to help them avoid infection.

Vaccines traditionally have been considered the most effective societal interventions to mitigate the impact of an epidemic of infectious disease such as influenza. After the initial cases are reported, and when the specific, causative strain had not been anticipated, a period of up to 6 months will elapse before immunization can be expected to protect members of the population from contracting the disease. During this time, the infectious agent needs to be characterized, and the vaccine must be configured, tested, manufactured, and distributed. Some additional days will be needed until the persons to whom the vaccine was administered develop immunity. Scientific and technological advances will someday reduce these time lags. Until then, or unless an efficacious "universal" flu vaccine is developed, the elapsed time between first cases and vaccine availability is, unfortunately, unavoidable.

Our flu case study contains two principal themes. The first is to examine the experience of vaccine allocation and distribution in the USA during the 2009 H1N1 influenza pandemic. We examine whether vaccine was administered in time materially to affect the course of the disease outbreak. The second theme is to make use of OR approaches to frame the question of whether there is a better way to distribute and allocate vaccine on the basis of the "dynamics" of the disease, which might allow substantial quantities of vaccine to be sent to geographical regions where they could be expected to offer the greatest benefits to the population.

2.4 Vaccine Allocation and Distribution During the 2009 H1N1 Pandemic

First, let us look at how vaccine was actually distributed in the recent influenza outbreak, in relation to the outbreak of cases of illness. We compared the statewide case incidence of influenza over time to two sources of vaccine distribution data. In the USA, all vaccines are developed and allocated by one governing body, the CDC. Our first set of data is the vaccine shipment data, which track the number of vaccines shipped to each state over time. This information was provided on a weekly basis by the CDC during the initial vaccination period (Centers for Disease Control and Prevention 2010). We obtained these data for 50 states. The second source provided data on vaccine actually administered, as each health-care provider was required to report quantities to state health authorities before being given additional vaccine. These federally mandated vaccine administration tracking data were aggregated on a weekly basis and forwarded to the CDC (Centers for Disease Control and Prevention 2012). We obtained this information from individual state health departments for 11 states (Finkelstein et al. 2011a).

Accounting for an 8–10 day delay once an individual receives vaccine before the development of immunity (Washington State Department of Health 2010), and observing from our data a delay of at least a week between vaccine shipment and delivering vaccinations into people's arms, some 2 weeks likely would elapse from the first week of vaccine shipment (Week of October 10, MMWR Week 40) before the first vaccinated members of a state's population would be protected from contracting the illness. We observed that in 24 of the 50 states the epidemic had already begun to decline before any individuals receiving the vaccine likely would have been protected from contracting the infection. Further, among the 11 states for which we were able to obtain data on actual vaccines administered, no more than 2% of the population received vaccinations before the outbreak had peaked. Consider Fig. 2.1, where we illustrate the epidemic curves of the USA as a whole together with two states with very different timing of H1N1 epidemic curves.

Note that while Maine received its vaccines in time to prepare for the oncoming epidemic, Georgia received its vaccines well after the peak. Figure 2.2 summarizes the "early/late" situation for all 50 US states.

After an outbreak of contagious illness has peaked, it is, of course, still possible to contract it, but much less likely. So, vaccinating population members would be expected to continue to offer some, albeit declining, benefit. However, the decline in the number of new cases is synonymous with the observation that "herd immunity" has been achieved, at which time the risks of transmitting disease decline rapidly.

Both the disease occurrence data and the vaccine availability data are subject to limitations. In the USA, the data collection channel from individual health-care providers and institutions through state health departments to the CDC suffers from chronic underreporting. On the other hand, some speculate that the "worried well" report to hospital emergency rooms, prompted by media coverage, leading to

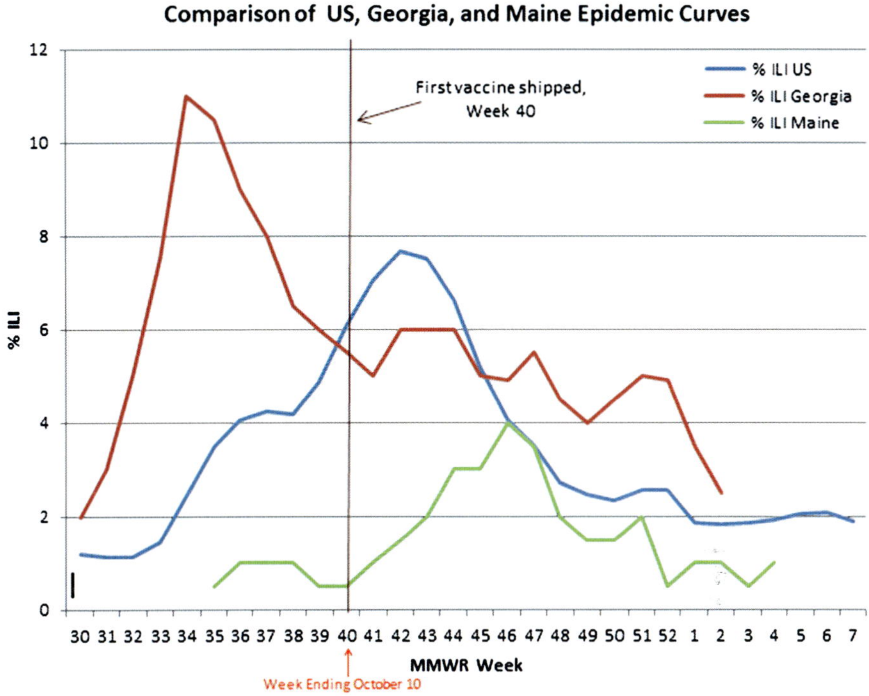

Fig. 2.1 Comparison of USA, Georgia, and Maine epidemic curves

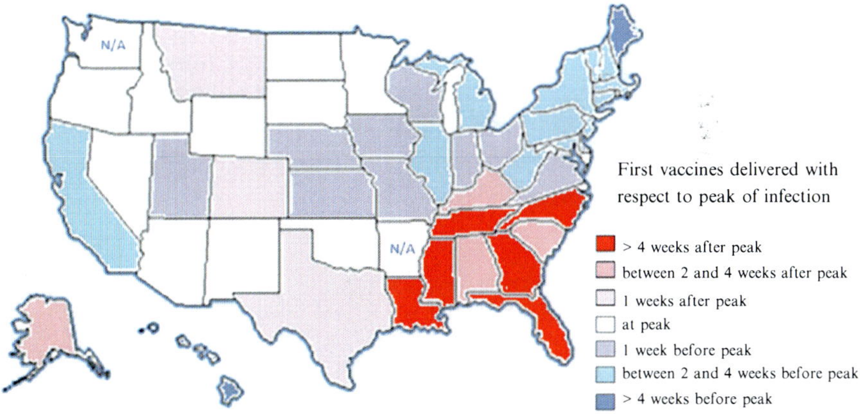

Fig. 2.2 Timing of shipment of first vaccine with respect to the peak of the infection in US states

overestimates of true influenza-like illness (ILI) incidence. Despite these possible complications, we believe that the available data support the broader observation that relatively few individuals had the opportunity to be vaccine-protected until after the risks of acquiring the flu from others were dramatically reduced.

2.5 Steps to Take Prior to Arrival of the Vaccine

The data we present characterize a situation in which vaccine was not available until well after influenza incidence peaked and began to decline in half of the states. In the absence of faster and more reliable vaccine production methods, it is likely that such a situation will occur again. Our results reinforce the question: what can public health officials do to reduce the risk of contracting flu before vaccine is available? What can individuals and families do?

A sensible approach, prior to arrival of vaccine, would be to focus on Non-pharmaceutical Interventions (NPIs), especially diligent hygienic behavior and reduction of human-to-human contacts. Until scientific and technological advances serve to reduce significantly the time needed to develop, test, produce, and administer vaccine, NPIs may be the only viable recourse to protect the millions of at-risk individuals.

To understand the potential very positive effects of NPIs, we need to discuss briefly an equation that relates human behavior to disease spread. As anyone who has seen the film *Contagion* (Contagion 2011) knows, the fundamental parameter of disease progression is R_0, where

$R_0 =$ the mean number of new infections caused by a newly infected person at the start of the epidemic when nearly everyone is susceptible to the disease.

To be an epidemic, we must see $R_0 > 1.0$. That is, each newly infected person will—on average–infect more than one additional person at the start of the epidemic, thereby creating exponential increases in the numbers infected. If $R_0 < 1$, then there is no exponential increase and no epidemic; the disease simply dies away.

Here is the good news: While R_0 does depend on characteristics of the particular flu virus, it also depends strongly on the behavior of humans. Close proximity and lack of good hygienic behavior can dramatically increase R_0. Likewise, reducing proximity to others by "social distancing" and being diligent in hygiene (such as vigorous hand washing several times a day) can dramatically reduce R_0. Here is the OR equation that summarizes these relationships:

$$R_0 = \lambda p,$$

where

$\lambda =$ average number of daily face-to-face human contacts and $p =$ "transmission probability" $=$ conditional probability that an infectious person will pass on the infection during a random face-to-face contact with a susceptible person (Larson and Nigmatulina 2010).

We can control λ by reducing our number of daily face-to-face human contacts. And we can reduce p by hand washing; not touching our mouth, nose, and eyes; not shaking hands with people (perhaps substituting the "elbow bump"!), and more. In a typical flu situation, R_0 is usually in the range of 1.2–1.8. Only a 30% reduction in each λ and p can bring R_0 down to below 1.0! If everyone did that, we could—in theory—by human NPI behavior alone stop the spread of the flu. Even if we cannot accomplish that goal, we can dramatically reduce R_0, thereby reducing the chances that we and our loved ones will become infected. This is the best we can do prior to the arrival of the vaccine. Additional NPI strategies are discussed in two recently published papers (Finkelstein et al. 2009, 2011b).

2.6 Using Operations Research Approaches to Find a Better Way to Allocate Vaccine

We now use OR approaches to examine what went wrong with vaccine allocation and distribution during the 2009 H1N1 pandemic. We fit a model to this disease occurrence, generate a similar model assuming no vaccine, and then estimate the numbers of averted cases of illness. This information then informs our decisions about deploying available vaccine in the future.

To estimate the number of infections averted from various vaccine programs, we first use available data to estimate the epidemic curve during the period in the fall of 2009 when H1N1 was most prevalent. Once we obtain an estimated epidemic curve, we use it in conjunction with reported vaccine administration data to fit the observed epidemic curve to the one generated by a mathematical model based on difference equations. We use a discrete-time version of the standard Kermack–McKendrick model to estimate infection spread within each state (McKendrick and Kermack 1927). In the model calibration process, we estimate the relevant parameters such as R_0 within each state. We estimate the R_0 for each state individually, since different states have different demographic, geographic, and cultural attributes. Moreover, states experienced the H1N1 outbreak at different times and implemented their vaccination programs in different ways (Hopkins 2009), so the extent of infection varied markedly. We then estimate a different, *non-observable* flu wave curve for the scenario in which no vaccine is available. How do we do that? We use the same model, just calibrated to reported data, but remove the vaccine component of the model. That is, we obtain an estimate of what would have happened if no vaccine had been delivered. This multi-step process provides a data-informed, model-supported basis for estimating the positive effects, if any, of the vaccine as administered in each of the states.

We refer the reader to Larson and Teytelman (2011) for the detailed discrete-time equations.

2.7 Results

Consider Oklahoma as an illustrative example of the modeling process. Figure 2.3 contains (1) the epidemic curve for the US state of Oklahoma and (2) the time-sequenced vaccine administration data reported by the state.

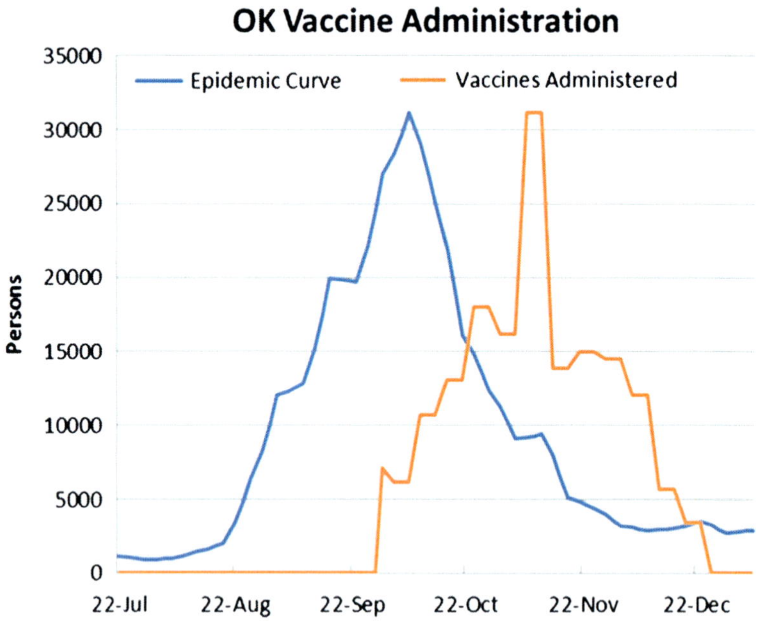

Fig. 2.3 Estimated Oklahoma epidemic curve compared to vaccines as administered in the state

In Fig. 2.4 we again include the empirical epidemic curve and three model-generated epidemic curves:

- The curve generated by using the Oklahoma-reported vaccine administration data fitted to correspond best to the empirical epidemic curve.
- The curve generated in the hypothetical case where vaccines were not administered at all.
- The curve generated in the hypothetical case where vaccines were administered 2 weeks earlier than had actually occurred.

The total number of infections in Oklahoma is calculated by finding the area under an epidemic curve. The estimated effect of the vaccines administered in Oklahoma can be determined by calculating the area between the "actual" model-generated curve, and the "no-vaccine" model-generated curve (Fig. 2.5).

We analyzed 11 states in the same manner and inferred the total number of infections that have been prevented as a result of their respective immunization programs. Officials in Illinois, Indiana, Massachusetts, Mississippi, Montana, New Jersey, New York, North Dakota, Oklahoma, South Carolina, and Virginia graciously provided us with precise data on vaccines as they were dispensed throughout the outbreak. In Table 2.1 we display two cases for each state:

1. The optimistic case, in which all vaccines are effective immediately and are 100% effective.

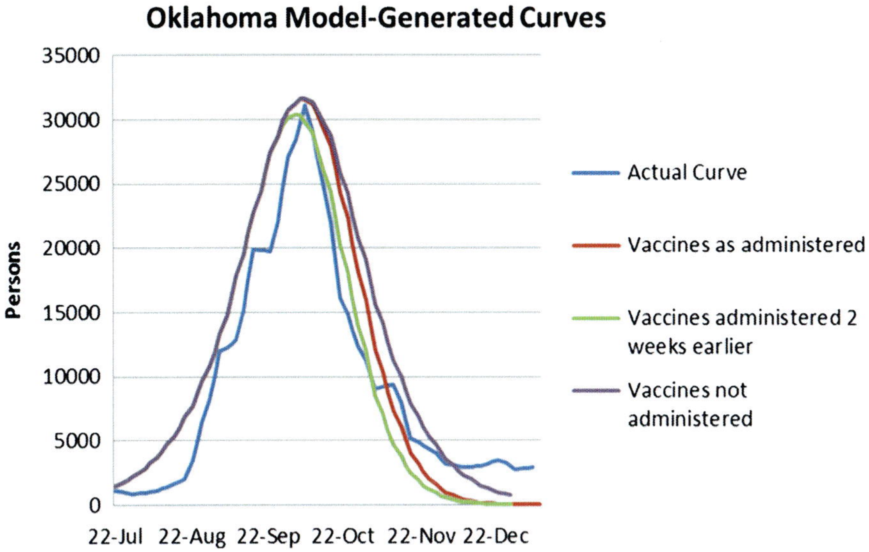

Fig. 2.4 The estimated epidemic curve along with the model-generated curves with and without vaccines

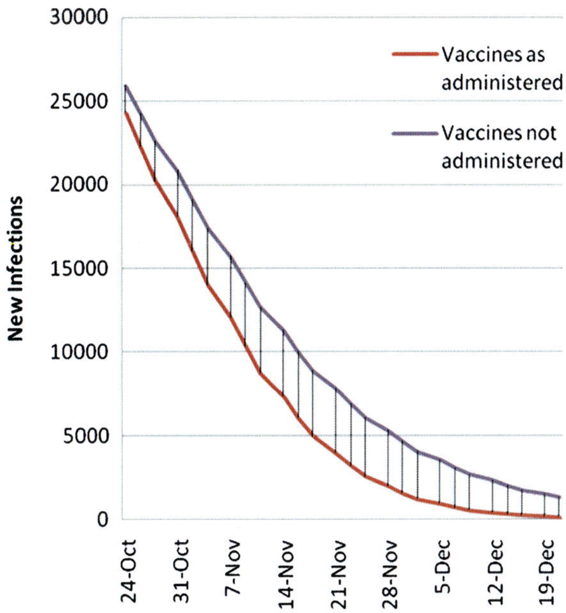

Fig. 2.5 A closer look at Fig. 2.4. (The *shaded* region represents the difference between the estimated number of H1N1 infections that would have occurred without intervention of vaccines and the estimated number of infections that actually occurred with the vaccine)

Table 2.1 Summary of best-fitted values for R_0 and model-determined effects of vaccines as they were distributed

(3) State	Estimated R_0	% Population vaccinated	Optimistic infections averted (in % of total population)	Pessimistic infections averted (in % of total population)
Illinois	1.21	9	3.43	1.2
Indiana	1.15	20	4.28	1.81
Massachusetts	1.16	29	13.71	6.84
Mississippi	1.16	8	0.13	0.05
Montana	1.15	20	2.81	1.04
New Jersey	1.20	12	3.36	1.1
New York	1.20	14	3.23	1.12
North Dakota	1.16	27	2.95	1.06
Oklahoma	1.14	13	2.29	0.93
South Carolina	1.16	8	0.4	0.12
Virginia	1.19	22	1.77	0.52

In the optimistic scenario vaccines are 100% effective and take effect immediately. In the pessimistic scenario vaccines are 75% effective and take effect 2 weeks later

2. The pessimistic case, in which vaccines are effective 2 weeks after administration and are effective for only 75% of the individuals receiving the vaccine.

The actual effect of the vaccine should lie within the range specified by these two cases.

2.8 Discussion

While examining the estimated number of infections averted, we can identify two major contributing factors. The first is the total number of vaccines administered to the general population. The second is the timing of the vaccine administration with respect to the peak of the infections. Once the H1N1 virus had been identified as a potentially devastating pandemic in the spring of 2009, the CDC worked to develop and distribute H1N1 vaccines. These vaccines were sent out to the individual states at the same time, and first doses were administered on October 5, 2009. These vaccines, however, had varying effects as the peak of the outbreak in different states occurred at markedly different times.

The peak of infection usually occurs when "herd immunity" occurs, which is the time when every infectious person at that time infects on average just one other person. At this point, R_0 no longer applies, as many of people have recovered from the disease and are now immune to further infection. The "real time" R_0 at time of herd immunity, called $R(t_{\mathrm{HI}})$, where t_{HI} is the time of herd immunity, is equal to 1.0. At the time of herd immunity, the number of infections in the next generation is approximately the same as it was in the previous one. We say "approximately" due to statistical fluctuations in the actual number of susceptible people that any one newly infectious person will infect. Soon afterwards, infectious people no longer replace themselves with newly infected individuals, and so the number of infected and infectious people in each generation decreases. Earlier administration of vaccines decreases the number of people who still "need to be infected" for the population to reach herd immunity and decreases the height of the peak. Late administration of the vaccine has almost no effect on the dynamics of the outbreak, and has little benefit to the society other than immunizing the people who received the vaccine. Such late immunizations may be important if the flu were to return later in a new wave.

Consider again the southeastern states of the USA, the first region to report infection peaks. As early victims they received vaccines after the worst of the infection had already passed. Louisiana, Indiana, and South Carolina did not start administering vaccines until *after* the peak of the outbreak. And these states were least successful in averting infections. On the other hand, Massachusetts and Virginia started administering their vaccines 5 and 3 weeks, respectively, before their respective peaks. These two states enjoyed a particularly good impact from their vaccination programs. In addition to having 5 weeks of vaccinations prior to the H1N1 peak in Massachusetts, that state vaccinated 29% of its population, the most of any state in our sample. As a result, as much as 7–14% of the population may have been spared infection and possible complications from influenza. While Massachusetts and Virginia had effective experiences with their vaccinations, most states did not. On average for our sample, vaccines were delivered just before the peak of the states' outbreaks.

To quantify the effect of time in averting infection we consider one of the states that vaccinated almost 20% of its population, Indiana. The hypothetical case where the same number of vaccines is delivered just 2 weeks earlier resulted in averting more than twice the number of infections than the vaccines as administered. Timing is everything!

With a more granular approach, we considered the marginal benefit of administering just one vaccination at a given time. We mapped the total projected number of infections that could be averted if just one vaccination were to be administered to a random susceptible person at different times during the outbreak. That is, we calculated the total number of infections that would occur in Indiana if exactly one vaccine were administered at different points in time, and compared that number to the total number of infections that would happen if no vaccines were administered at all. The differences are presented in Fig. 2.6. As expected, administered vaccines have a monotonically decreasing benefit with respect to time. A striking feature of Fig. 2.6 is the fact that one vaccination to a susceptible person well before the flu wave starts averts almost two infections in the population, even with a low value for R_0 (1.15) and even considering the fact that the vaccinated person has a greater than ever chance of never becoming infected assuming no vaccination. Clearly, vaccines administered well before the peak carry the added benefit of diluting the susceptible population with immune people and are particularly useful in mitigating the spread of infection.

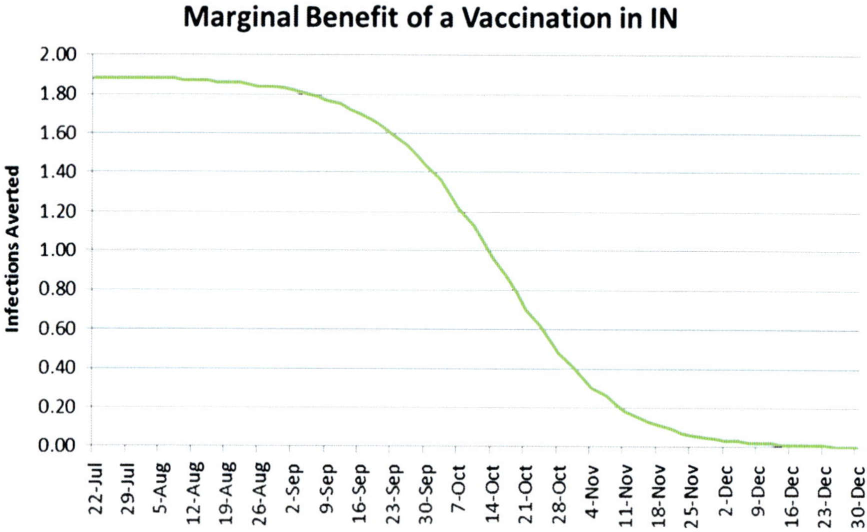

Fig. 2.6 The graph shows the number of infections averted by administering exactly one vaccination to a susceptible person at different points throughout the outbreak

Another insightful feature of the graph of Fig. 2.6 is the slope of the marginal benefit curve, which represents the time-dependence of effective vaccines. While starting vaccine administration in Indiana in July would be most effective for Indiana, the effect of those vaccines would not change significantly until the beginning of September. That is, if vaccines were to be available in July, Indiana could have waited to receive its share until September with minimal losses. Similarly, vaccines received after December will have the same (minimal) effect whether they are administered in December or February. The effectiveness of vaccines, however, is extremely time-sensitive from the end of September to mid-November, where each week results in a significant loss of effectiveness. Vaccines that become available during this critical period need to be administered as soon as possible. These results encourage us to recommend a more detailed cost–benefit analysis of trying to get some vaccine, even if in much smaller quantities, to the states at the beginning of this "critical period," when the population is particularly sensitive to the timing of vaccination. A small amount of vaccine delivered early should have a more significant effect on the total number of infections averted than a batch delivered just a few weeks later.

The timeliness of the vaccines also appears to be closely related to the total amount of vaccine accepted by the population. Largely dictated by political considerations, the CDC distributes its vaccines proportionally according to the population of each region. This allocation process does not minimize the total number of infections incurred nationally. Once the regional peak of the outbreak had passed and H1N1 had been determined to be less dangerous than originally feared, the populations of the "early victim" states were less likely to obtain a flu vaccination. Their decision could relate to time spent getting the flu shot and perceived risks' possible side effects. States vary considerably in the amount of vaccine that was actually used. Mississippi used less than 40% of its allocated vaccine, most likely due to "flu fatigue." While the media are particularly helpful at warning the public of an ongoing pandemic and encouraging diligent hygienic behavior and social distancing through school closures and cancellation of public events, they can also give the impression that the outbreak is over or has been blown out of proportion—thereby creating flu fatigue.

2.9 Thought Experiment

As shown in Fig. 2.7, in the first few weeks of vaccine distribution, when demand for vaccine clearly exceeded supply, the CDC allocated vaccine to states proportionally to their populations (Centers for Disease Control and Prevention 2010). Particularly in early October, this simple distribution scheme ensured that all states received amounts that could be used to immunize the same proportion of the population. Come November, those states that saw little demand started placing fewer orders for vaccine, while those with later epidemics like Massachusetts and Virginia were still experiencing high demand and were shipped larger quantities of vaccine, confirming the intuition from Fig. 2.6.

Thus we see that the same vaccines administered in states that already experienced the peak of the infection at the time vaccines started arriving were much less effective than those administered in states that had not yet experienced the peak.

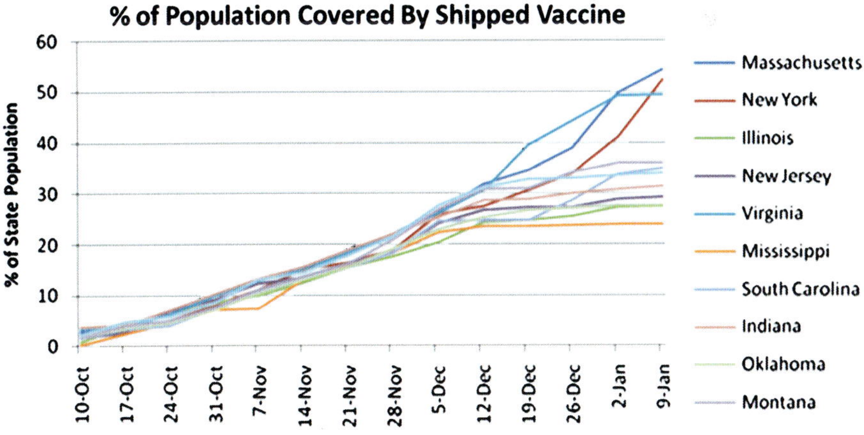

Fig. 2.7 CDC distribution of vaccines, initially approximately proportional to state populations

Table 2.2 Comparison of vaccination programs in North Dakota and Mississippi

North Dakota	Mississippi
• Started administration 1 week before peak	• Received vaccines 6 weeks after peak
• A hypothetical single batch of 32,342 vaccines (5% of the population) administered on October 10, averts 33,745 infections (5.2% of the population)	• A hypothetical single batch of 147,599 vaccines (5% of the population) administered on October 10, averts 27,173 infections (0.9% of the population)
• In the first 4 weeks administered 62% of available vaccines	• In the first 4 weeks administered less than 25% of available vaccines

States that were past the peak saw less demand for vaccines and thus used only a fraction of their allocated vaccine. Consider a side-by-side analysis of Mississippi and North Dakota, as shown in Table 2.2.

It is clear that vaccines administered in North Dakota were significantly more effective than those in Mississippi. In fact, the Mississippi vaccines had almost no effect because the infection was barely spreading by the time vaccines became available. Coupled with this, North Dakota was experiencing more demand for the vaccines at the beginning of its program. Motivated by this we believe that there is a need for more effective procedures of allocating vaccines to US States.

Allocating all of Mississippi's vaccine to North Dakota would be not only unethical but also politically infeasible. Instead, as a thought experiment, suppose that just 20% of Mississippi's *unused* vaccine were to be transferred to North Dakota during the first 4 weeks of vaccine distribution. Suppose that, with this addition, 60% of the new vaccines were actually administered. This additional vaccine would decrease the total number of infections in North Dakota by 5%. That is a significant improvement for a relatively small cost. An adaptive decision like this can be made during the allocation process. We can form even approximate predictions about how much vaccine will actually be demanded by the state and how effective the extra vaccines would be in reducing infections. For example, by using data collected from our 11 states, we can weakly estimate that a state that experiences the peak of infection 6 weeks before receiving the first shipment of vaccine can be expected to vaccinate no more than 4% of its population in the first 4 weeks of the vaccination program. In the first 4 weeks of the 2009 H1N1 pandemic, the CDC allocated to Mississippi enough vaccine to cover 7% of its population. With accurate data, some portion of that could have been redirected to states that are more likely to use and benefit from the vaccine.

Flu-specific takeaways:

- Early vaccines are significantly more effective than late vaccines.
- Data shows that vaccines that start being administered early on in the epidemic tend to have a higher acceptance rate in the population.

- There exists a time-sensitive "critical period" within a few weeks of the peak of an epidemic where the effectiveness of the vaccines is extremely time-dependent. This is the time when it is important to vaccinate as much of the population as possible.
- Practitioners taking these points into account and combining them with OR resource allocation methods will see significant improvements in the benefits of available vaccine.

2.10 Conclusion

The use of OR methods to address homeland security issues, such as an influenza pandemic, forces a logical and systematic examination of the whole problem, rather than just certain parts of it. Commonly, the result is to uncover relationships that are non-intuitive, or even counter-intuitive. An improved understanding of the whole problem can benefit planning at strategic, tactical, and operational levels.

Let us again consider the context of the allocation of vaccine by the CDC to the individual states once a pandemic is already underway. If and when a southern state, for example, experiences a wave of cases, a natural tendency would be to make a tactical decision to ship via express courier the state's proportional share of the vaccine to address what appears to be an imminent, growing problem. But the total system situation is far more complicated than a one-state-at-a-time decision analysis. Consider *Scenario* 1: The southern states are just now experiencing a rapidly rising flu wave and the northern states do not yet have the flu. In this situation, the southern states should receive priority and be shipped a number of vaccines above their proportional population share. This is because the northern states can wait without significant penalty. Now consider *Scenario* 2: The vaccines are delayed and the southern states have all reached peak, herd immunity has occurred, and their respective flu waves have started to decrease. By the time the vaccine reaches the southern state in question, many who were sick would have recovered, are now immune, and risks to remaining healthy persons have become much less. A better decision would be to increase the allocation to more vulnerable northern regions that had so far seen few cases and must have many more susceptible individuals.

The results of one OR analysis might suggest the need for others. For example, it is highly likely that a time interval will exist after cases of illness are recognized, but before vaccine can be produced and distributed. The value of diligent hygiene and social distancing, especially under these circumstances, are already widely recognized. In planning for future outbreaks of illness, it would be useful to quantify the benefits of devoting resources to systematic campaigns to modify certain patterns of personal behavior, in relation to the effort's costs.

A great benefit of operations research is that it forces a logical and systematic consideration of all aspects of a problem. Pandemics and other LPHC events gravely threaten lives and the security of our homeland. We are hopeful that the approaches we have described offer the prospect of mitigating the future impact of these kinds of adverse events.

Acknowledgments Work on this chapter was supported by the Sloan Foundation of New York under a grant entitled *Decision-Oriented Analysis of Pandemic Flu Preparedness & Response* and under a cooperative agreement with the US Centers for Disease Control and Prevention (CDC), grant number 1 PO1 TP000307-01, *LAMPS* (*Linking Assessment and Measurement to Performance in PHEP Systems*), awarded to the Harvard School of Public Health Center for Public Health Preparedness (HSPHCPHP) and the Massachusetts Institute of Technology (MIT) Center for Engineering Systems Fundamentals (CESF). The discussion and conclusions in this chapter are those of the authors and do not necessarily represent the views of the Sloan Foundation, the CDC, the US Department of Health and Human Services, Harvard, or MIT.

References

Centers for Disease Control and Prevention (2010) Graph and table of 2009 H1N1 influenza vaccine doses allocated, ordered, and shipped. http://www.cdc.gov/h1n1flu/vaccination/vaccinesupply. Accessed 2 Feb 2010

Centers for Disease Control and Prevention. Public Health Information Network, Countermeasure and Response Administration, Novel Influenza A (H1N1) Response. http://www.cdc.gov/phin/tools/cra/doses_administered.html. Accessed 2 Oct 2012

Contagion (2011). http://contagionmovie.warnerbros.com/index.html. Accessed 18 Nov 2011

Eshghi K, Larson RC (2008) Disasters: lessons from the past 105 years. Disaster Prev Manage 17(1):62–82

Finkelstein S, Prakash S, Nigmatulina K, Klaiman T, Larson RC (2009) Pandemic influenza: non-pharmaceutical interventions and behavioral changes that may save lives. Int J Health Manage Inf 1(1):1–18

Finkelstein SN, Hedberg KJ, Hopkins JA, Hashmi S, Larson RC (2011a) Vaccine availability in the United States during the 2009 H1N1 outbreak. Am J Disaster Med 6(1):23–30

Finkelstein S, Prakash S, Nigmatulina K, McDevitt LRC (2011b) A home toolkit for primary prevention of influenza by individuals and families. Disaster Med Public Health Prep 5:266–271

Greenfieldboyce N (2011) Bird flu research rattles bioterrorism field. http://www.npr.org/blogs/health/2011/11/17/142453447/bird-flu-research-rattles-bioterrorism-field. Accessed 17 Nov 2011

Hopkins J (2009) 2009 H1N1 after action reports: lessons on vaccine distribution. MIT ESD Working Paper. http://esd.mit.edu/staging/WPS/2011/esd-wp-2011-06.pdf. Accessed 2 Oct 2012

Koopman BO (1957) The theory of search. III. The optimum distribution of searching effort. Oper Res 5(5):613–626

Larson RC (1972) Improving the effectiveness of New York City's 911. In: Drake AW, Keeney RL, Morse PM (eds) Analysis of public systems. MIT Press, Cambridge

Larson RC (2005) Decision models for emergency response planning. In: Kamien D (ed) The McGraw-Hill handbook of homeland security. McGraw-Hill, Columbus, Ohio, pp 911–927

Larson RC, Nigmatulina KR (2010) Engineering responses to pandemics. In: Rouse W, Cortese D (eds) Engineering the system of healthcare delivery. IOS Press BV, Amsterdam

Larson RC, Teytelman A (2011) Modeling the effects of H1N1 influenza vaccine distribution in the United States. Value Health 15(1):158–166

Larson RC, Metzger MD, Cahn MF (2006) Responding to emergencies: lessons learned and the need for analysis. Interfaces 36(6):486–501

Leonhardt D (2004) Incremental analysis, with two yards to go. http://www.math.toronto.edu/mpugh/Teaching/Sci199_03/Football_game_theory.htm. Accessed 7 Nov 2011

McKendrick AG, Kermack WO (1927) Contributions to the mathematical theory of epidemics. Proc R Soc Lond A 115:700–721

Morse PM, Kimball GE (2003) Methods of operations research. Courier Dover Publications, Mineola

Washington State Department of Health (2010). H1N1 Swine Flu, Vaccine, WA State Dept. of Health. http://www.doh.wa.gov/h1n1/h1n1_vaccine.htm. Accessed 17 June 2010

Chapter 3
Deployed Security Games for Patrol Planning

Fernando Ordóñez, Milind Tambe, Juan F. Jara, Manish Jain, Christopher Kiekintveld, and Jason Tsai

Abstract Nations and organizations need to secure locations of economic, military, or political importance from groups or individuals that can cause harm. The fact that there are limited security resources prevents complete security coverage, which allows adversaries to observe and exploit patterns in patrolling or monitoring and enables them to plan attacks that avoid existing patrols. The use of randomized security policies that are more difficult for adversaries to predict and exploit can counter their surveillance capabilities and improve security. In this chapter we describe the recent development of models to assist security forces in randomizing their patrols and their deployment in real applications. The systems deployed are based on fast algorithms for solving large instances of Bayesian Stackelberg games that capture the interaction between security forces and adversaries. Here we describe a generic mathematical formulation of these models, present some of the results that have allowed these systems to be deployed in practice, and outline remaining future challenges. We discuss the deployment of these systems in two real-world security applications: (1) The police at the Los Angeles International Airport uses these models to randomize the placement of checkpoints on roads entering the airport and the routes of canine unit patrols within the airport terminals. (2) The Federal Air Marshal Service (FAMS) uses these models to randomize the schedules of air marshals on international flights.

F. Ordóñez (✉) • J.F. Jara
Industrial Engineering Department, University of Chile, Republica 701, Santiago, Chile
e-mail: fordon@dii.uchile.cl

M. Tambe • M. Jain • J. Tsai
Computer Science Department, University of Southern California, Los Angeles, CA 90089, USA
e-mail: milind@usc.edu; manishja@usc.edu; jasontts@usc.edu

C. Kiekintveld
Computer Science Department, University of Texas, El Paso, TX 79968, USA
e-mail: cdkiekintveld@utep.edu

J.W. Herrmann (ed.), *Handbook of Operations Research for Homeland Security*, International Series in Operations Research & Management Science 183, DOI 10.1007/978-1-4614-5278-2_3,

3.1 Introduction

Nations and organizations need to secure locations of economic, military, or political importance from groups or individuals that can cause harm. Protecting such critical sites and targets, such as airports, historical landmarks, power generation facilities, and political figures, is a challenging task for police and security agencies worldwide. The growing threat of international terrorism has exacerbated this challenge in recent years. For instance, transportation networks such as buses, trains, and airplanes carry millions of people per day to their destinations, making them a prime target for terrorists and extremely difficult to protect for law enforcement agencies. The September 11, 2001 attack on the World Trade Center in New York City via commercial airliners resulted in $27.2 billion of direct short-term costs (Looney, 2002) as well as a government-reported 2,974 lives lost. The 2004 Madrid commuter train bombings resulted in 191 lives lost, 1,755 wounded, and an estimated cost of 212 million Euros (Blanco et al., 2007). Finally, in the 2005 London subway and bus bombings, 52 lives were lost, 700 were wounded, and there was an estimated economic cost of two billion pounds (Thornton, 2005).

Measures for protecting potential target areas include monitoring entrances or inbound roads and patrolling the network at transfer points and aboard transportation vehicles. However, limited resources imply that it is typically impossible to provide full security coverage at all times. Furthermore, adversaries can observe security arrangements over time and exploit any predictable patterns to their advantage. One way to mitigate the ability of adversaries to exploit patterns is the judicious use of randomization in scheduling the actions of security forces. For example, police patrols, baggage screenings, vehicle checkpoints, and other security procedures are often randomized. However, security forces face many difficulties in effectively randomizing their operations. One of these difficulties is how to weigh the different actions the defender could take. A strategy in which all targets are equally likely to be defended fails to take into account that some targets are more attractive or vulnerable than others. A defense strategy that weighs the protection of each target against the value of that target still fails to account for the possibility that the attacker is intelligent and will update their strategy based on the actions of the defender. Asking a human to generate a random security policy has additional difficulties as humans are not good at generating truly random behavior (Wagenaar, 1972; Treisman and Faulkner, 1987) and can easily fall into predictable patterns. Furthermore, in transportation networks and many other security domains, the problem of scheduling security forces is prohibitively large, even without considering randomization. Creating a schedule by hand is a costly and labor-intensive process.

Our work on randomized patrol planning has lead to a number of deployed software assistants that address many of these key difficulties of randomization and provide an easy-to-use solution for security forces. These assistants use game-theoretic models and solution algorithms to determine good randomization strategies that take into account target values and assume intelligent adversary responses to security measures. Game theory is a well-established paradigm for reasoning about situations with multiple self-interested decision makers (Fudenberg and Tirole, 1991). We model security games as Stackelberg games (von Stackelberg, 1934) between the defender (i.e., the security forces) and the attacker (i.e., a terrorist adversary). Stackelberg games are a bilevel model (Bard, 1999) that account for the ability of an attacker to gather information about the defense strategy before planning an attack. These games specify different payoff values for both players in the event of an attack on every potential target. Extending these games to Bayesian Stackelberg games (Conitzer and Sandholm, 2006) allows us to capture uncertainty about these payoffs in the game model. Solutions to these games provide a randomized policy for the defense strategy, which can be used to generate specific schedules for security patrols.

In this chapter we describe how we applied this game-theoretic approach in two different software solutions that provide assistance in scheduling real security operations. The ARMOR program (Pita et al., 2008), developed for the Los Angeles international airport (LAX) police, randomizes checkpoints on the roadways entering the airport and canine patrol routes within the airport terminals. The IRIS program (Tsai et al., 2009) was developed for the Federal Air Marshal Service (FAMS) to assist with randomly scheduling air marshals on flights. These software assistants are interactive, and domain experts can change domain parameters when necessary. Underlying each of these tools is a model of the domain as a Bayesian Stackelberg game, along with fast solution algorithms for computing an optimal solution to the game model. These algorithms use various techniques for exploiting structure in the security domains to speed up the computation and enable large real-world problem instances to be solved in reasonable amounts of time (Paruchuri et al., 2006, 2007; Kiekintveld et al., 2009). As we highlight later in this chapter, developing these assistants requires a substantial amount of work in calibrating the Stackelberg game model to capture expert's knowledge of the security domain. This is critical so that the defender strategies proposed are reasonable and useful. Clustering and data-mining methods can help in formulating a representative security game in situations where there is sufficient information of events.

The rest of the chapter is organized as follows. Related work is discussed in Sect. 3.2. The Bayesian Stackelberg security game models, technical formulation, and solution algorithms are discussed in Sect. 3.3. The LAX and FAMS domains and the software assistants developed are described in Sect. 3.4. In Sect. 3.5, we illustrate how to use clustering methods to automatically build a Stackelberg security game for a network patrolling problem. We present our conclusions in Sect. 3.6. This chapter is based on our previous work (Paruchuri et al., 2007; Kiekintveld et al., 2009; Jain et al., 2010) and extends it by describing a data-driven process to build a Stackelberg security game.

3.2 Related Work

There are three main areas of related work that we review here: Optimization techniques for patrol planning that do not take the strategic behavior of adversaries into account, Stackelberg game models used in diverse security problems, and other game-theoretic models used for security.

The first area of related work applies optimization techniques that model a security domain but do not address the strategic aspects of the problem. These methods provide a randomization strategy for the defender, but they do not take into account the fact that the adversaries can observe the defender's actions and then adjust their behavior. Examples of such authors or approaches include Ruan et al. (2005) and Paruchuri et al. (2006), which are based on learning, Markov decision processes (MDPs) and partially observable Markov decision processes (POMDPs). As part of this work, the authors model the patrolling problem with locations and varying incident rates in each of the locations and solve for optimal routes using a MDP framework. Another example is the "Hypercube Queueing Model" (Larson, 1974) which is based on queueing theory and depicts the detailed spatial operation of urban police departments and emergency medical services. It has found application in police beat design, in allocation of patrolling time, etc. Such frameworks can address many of the problems we raise, including different target values and increasing uncertainty by using many possible patrol routes. However, they fail to account for the possibility that an intelligent attacker will observe and exploit patterns in the security policy. If a policy is based on the historical frequency of attacks, it is essentially a reactive policy and an intelligent attacker will always be one step ahead.

A second area of work uses Stackelberg games to model a variety of security domains. Bier (2007) give a strong endorsement of this type of modeling for security problems. Game-theoretic models have been applied in a variety of homeland security settings, such as protecting critical infrastructure (Brown et al., 2006; Pita et al., 2008; Nie et al., 2007). Wein (2009) apply Stackelberg games in the context of screening visitors entering the USA. In their work, they model the US Government as the leader who specifies the biometric identification strategy to maximize the detection probability using finger print matches, and the follower is the terrorist who can manipulate the image quality of the finger print. They have also been used for studying missile defense systems (Brown et al., 2005) and for studying the development of an adversary's weapon systems (Brown et al., 2005). A family of Stackelberg games known as inspection games is closely related to the security games we are interested in and includes models of arms inspections and border patrols (Avenhaus et al., 2002). Other recent work uses Stackelberg games to obtain randomized patrolling in a generic "police and robbers" scenario (Gatti, 2008) and perimeter patrols (Agmon et al., 2008).

Our work belongs to this line of research, focusing on Stackelberg games for patrol planning. Our work differs from the previous work mainly in the solution approach used and the domain constraints considered, which have arisen from our work deploying these systems in the real world. In addition to the ARMOR and IRIS systems that will be discussed in detail in this chapter, we are currently working on designing new game-theoretic scheduling assistants for other security agencies. For instance, GUARDS (Pita et al., 2011) is a system for scheduling activities being developed for the Transportation Security Administration. GUARDS is being evaluated at an undisclosed airport for potential nationwide deployment. Finally, PROTECT (An et al., 2011) is in use for scheduling the patrols of the United States Coast Guard in the port of Boston; it is currently being deployed in the port of New York and may be deployed at multiple other ports in the USA.

The third area of related work is the application of game-theoretic techniques that are not based on Stackelberg games to security applications. Security problems are increasingly studied using game-theoretic analysis, ranging from computer network security (Lye and Wing, 2005; Srivastava et al., 2005) to terrorism (Sandler and Arce, 2003). Babu et al. (2006) have worked on modeling passenger security system at US airports using linear programming approaches; however, their objective is to classify the passengers in various groups and then screen them based on the group to which they belong.

3.3 Methodology

A generic Stackelberg game has two players, a *leader* and a *follower*. These players need not represent individuals but could also be groups that cooperate to execute a joint strategy, such as a police force or terrorist organization. For the modeling of security, the leadership role is assumed by the police, and the role of follower by criminals. Decisions made by each player are where to protect and where to attack respectively. Thus, having police act first reflects the fact that patrols conducted by police officers are observable by criminals and, in the long run, the latter are able to estimate the probability of encountering police in a given sector. Thus, the decision of offenders is carried out once the likelihood of facing the police is observed.

The actions for the security forces represent the action of scheduling a patrol or security procedure to protect a set of targets, e.g., a checkpoint at the LAX airport or assigning federal air marshals to a flight. The actions for an adversary represent possible attacks at one of the targets being protected, e.g., a terminal at LAX or a certain flight.

3.3.1 Stackelberg Equilibrium

In a Stackelberg game each player has a set of possible *pure strategies*, denoted $\sigma_d \in \Sigma_d$ and $\sigma_a \in \Sigma_a$. A *mixed strategy* allows a player to play a probability distribution over pure strategies, denoted $\delta_d \in \Delta_d$ and $\delta_a \in \Delta_a$. Payoffs for each player are defined over all possible joint pure-strategy outcomes: $\Omega_d : \Sigma_a \times \Sigma_d \rightarrow \mathcal{R}$ for the defender and similarly for each attacker. The payoff functions are extended to mixed strategies in the standard way by taking the expectation over pure-strategy outcomes. The follower can observe the leader's strategy and then act in a way to optimize its own payoffs. Formally, the attacker's strategy in a Stackelberg security game becomes a function that selects a strategy for each possible leader strategy: $F_a : \Delta_d \rightarrow \Delta_a$.

The most common solution concept in game theory is a *Nash equilibrium*, which is a profile of strategies for each player in which no player can gain by unilaterally changing to another strategy (Osbourne and Rubinstein, 1994). Stackelberg equilibrium is a refinement of Nash equilibrium specific to Stackelberg games. It is a form of subgame perfect equilibrium in which it requires that each player select the best-response in any subgame of the original game (where subgames correspond to partial sequences of actions). The effect is to eliminate equilibrium profiles that are supported by non-credible threats off the equilibrium path. Subgame perfection is a natural requirement, but it does not guarantee a unique solution in cases where the follower is indifferent among a set of strategies. The literature contains two forms of Stackelberg equilibria that identify unique outcomes, first proposed by Leitmann (1978), and typically called "strong" and "weak" after (Breton et al., 1988). The strong form assumes that the follower will always choose the optimal strategy for the leader in cases of indifference, while the weak form assumes that the follower will choose the worst strategy for the leader. Unlike the weak form, strong Stackelberg equilibria are known to exist in all Stackelberg games (Basar and Olsder, 1995). A standard argument suggests that the leader is often able to induce the favorable strong form by selecting a strategy arbitrarily close to the equilibrium which causes the follower to strictly prefer the desired strategy (von Stengel and Zamir, 2004). We adopt strong Stackelberg equilibrium (SSE) as our solution concept in part for these reasons but also because it is the most commonly used in related literature (Osbourne and Rubinstein, 1994; Conitzer and Sandholm, 2006; Paruchuri et al., 2008).

Definition 1 A set of strategies (δ_d, F_a) form a *SSE* if they satisfy the following:

1. The leader plays a best-response:
   $\Omega_d(\delta_d, F_a(\delta_d)) \geq \Omega_d(\delta_d', F_a(\delta_d'))\ \forall \delta_d' \in \Delta_d$.
2. The follower plays a best-response:
   $\Omega_a(\delta_d, F_a(\delta_d)) \geq \Omega_a(\delta_d, \delta_a)\ \forall \delta_d \in \Delta_d, \delta_a \in \Delta_a$.
3. The follower breaks ties optimally for the leader:
   $\Omega_d(\delta_d, F_d(\delta_d)) \geq \Omega_d(\delta_d, \delta_a)\ \forall \delta_d \in \Delta_d, \delta_a \in \Delta_a^*(\delta_d)$, where $\Delta_a^*(\delta_d)$ is the set of follower best-responses, as above.

Whether or not the Stackelberg leader benefits from the ability to commit depends on whether commitment to mixed strategies is allowed. Committing to a pure strategy can be either good or bad for the leader; for example, in the "Rock, Paper, and Scissors" game, forcing commitment to a pure strategy would guarantee a loss. However, it has been shown that the ability to commit to a mixed strategy always weakly increases the leader's payoffs in equilibrium profiles of the game (von Stengel and Zamir, 2004). In the context of a Stackelberg security game, a deterministic policy is a liability for the defender (the leader), but a credible randomized security policy is an advantage. Our model allows commitment to mixed strategies by the defender.

The Bayesian extension to the Stackelberg game allows for multiple types of players, with each type associated with its own payoff values. For the security games of interest in this chapter, we assume that there is only one leader type (e.g., only one police force), although there are multiple follower types (e.g., multiple adversary types trying to infiltrate security). The set of follower types is denoted by Γ. Each type γ is represented by a different payoff matrix. The leader does not know the follower's type. The goal is to *find the optimal mixed strategy* for the leader to commit to, given that each follower type will know the mixed strategy of the leader when choosing its own strategy. Payoffs for each type are defined over all possible joint pure-strategy outcomes: $\Omega_d : \Sigma_a^\Gamma \times \Sigma_d \rightarrow \mathscr{R}$ for the defender and similarly for each attacker type. The leader's best response is now a weighted best response to the followers' responses, where the weights are based on the probability of occurrence of each type. The strategy of each attacker type γ becomes: $F_{\mathrm{a}}^{\gamma} : \Delta_{\mathrm{d}} \rightarrow \Delta_{\mathrm{a}}^{\gamma}$, which still satisfies constraints 2 and 3 in Definition 1.

3.3.2 Security Game Representation

There are two major problems with using conventional methods to represent security games in normal form. First, many solution methods require the use of a Harsanyi transformation when dealing with Bayesian games (Harsanyi and Selten, 1972). The Harsanyi transformation converts a Bayesian game into a normal-form game, but the new game may be exponentially larger than the original Bayesian game. Our compact representation avoids this Harsanyi transformation, and instead we directly operate on the Bayesian game. Operating directly on the Bayesian representation is possible in our model because the evaluation of the leader strategy against a Harsanyi-transformed game matrix is equivalent to its evaluation against each of the game matrices for the individual follower types. (For more details, see the Appendix; a further detailed explanation appears in Paruchuri et al. (2008)). The second problem arises because the defender has many possible resources to schedule in the security policy. This can also lead to a combinatorial explosion in a standard normal-form representation. For example, if the leader has m resources to defend n entities, then normal-form representations model this problem as a single leader with $\binom{n}{m}$ rows, each row corresponding to a leader action of covering m targets with security resources. However, in our compact representation, the game representation would only include n rows, each row corresponding to whether the corresponding target was covered or not. Such a representation is equivalent to the normal form representation for the class of problems we address in this work (see Kiekintveld et al. 2009 for additional details). This compactness in

Table 3.1 Example payoffs for an attack on a target

	Covered	Uncovered
Defender	5	−20
Attacker	−10	30

our representation is possible because the payoffs for the leader in these games simply depend on whether the attacked target was covered or not, and not on what other targets were covered (or not covered). The representation we use here avoids both of these potential problems, using methods similar to other compact representations for games (Koller and Milch, 2003; Jiang and Leyton-Brown, 2006).

We now introduce our compact representation for security games. Let $T = \{t_1, \ldots, t_n\}$ be a set of *targets* that may be attacked, corresponding to pure strategies for the attacker. The defender has a set of resources available to *cover* these targets, $R = \{r_1, \ldots, r_m\}$ (e.g., in the FAMS domain, targets could be flights and resources could be federal air marshals). Associated with each target are four payoffs defining the possible outcomes for an attack on the target, as shown in Table 3.1. There are two cases, depending on whether or not the target is covered by the defender. The defender's payoff for an uncovered attack when facing an adversary of type γ is denoted $U_{\mathrm{d}}^{\gamma,u}(t)$, and for a covered attack $U_{\mathrm{d}}^{\gamma,c}(t)$. Similarly, $U_{\mathrm{a}}^{\gamma,u}(t)$ and $U_{\mathrm{a}}^{\gamma,c}(t)$ are the payoffs of the attacker.

A crucial feature of the model is that payoffs depend only on the target attacked, and whether or not it is covered by the defender. The payoffs do *not* depend on the remaining aspects of the schedule, such as whether any unattacked target is covered or which specific defense resource provides coverage. For example, if an adversary succeeds in attacking Terminal 1, the penalty for the defender is the same whether the defender was guarding Terminal 2 or 3. Therefore, from a payoff perspective, many resource allocations by the defender are identical. We exploit this by summarizing the payoff-relevant aspects of the defender's strategy in a *coverage* vector, C, that gives the probability that each target is covered, c_t. The analogous attack vector A^γ gives the probability of attacking a target by a follower of type γ. We restrict the attack vector for each follower type to attack a single target with probability 1. This is without loss of generality because a SSE solution still exists under this restriction (Paruchuri et al., 2008). Thus, the follower of type γ can choose any pure strategy $\sigma_{\mathrm{a}}^{\gamma} \in \Sigma_{\mathrm{a}}^{\gamma}$, that is, attack any one target from the set of targets.

The payoff for a defender when a specific target t is attacked by an adversary of type γ is given by $U_{\mathrm{d}}^{\gamma}(t, C)$ and is defined in (3.1). Thus, the expectation of $U_{\mathrm{d}}^{\gamma}(t, C)$ over t gives U_{d}^{γ}, which is the defender's expected payoff given coverage vector C when facing an adversary of type γ whose attack vector is A^γ. U_{d}^{γ} is defined in (3.2). The same notation applies for each follower type, replacing "d" with "a." Thus, $U_{\mathrm{a}}^{\gamma}(t, C)$ gives the payoff to the attacker when a target t is attacked by an adversary of type γ. We will see $U_{\mathrm{a}}^{\gamma}(t, C)$ and $U_{\mathrm{d}}^{\gamma}(t, C)$ used in the MILP discussed later. We also define the useful notion of the *attack set* in (3.3), $\Lambda^\gamma(C)$, which contains all targets that yield the maximum expected payoff for the attacker type γ given coverage C. This *attack set* is used by the adversary to break ties when calculating a SSE. Moreover, in these security games, exactly one adversary is attacking in one instance of the game; however, the adversary could be of any type and the defender does not know the type of the adversary faced.

$$U_d^\gamma(t, C) = c_t U_d^{\gamma,c}(t) + (1 - c_t) U_d^{\gamma,u}(t) \tag{3.1}$$

$$U_d^\gamma(C, A^\gamma) = \sum\nolimits_{t \in T} a_t^\gamma \cdot (c_t \cdot U_d^{\gamma,c}(t) + (1 - c_t) U_d^{\gamma,u}(t)) \tag{3.2}$$

$$\Lambda^\gamma(C) = \{t : U_a^\gamma(t, C) \geq U_a^\gamma(t', C) \forall t' \in T\}. \tag{3.3}$$

In an SSE, the attacker selects the target in the attack set with maximum payoff for the defender. Let t^* denote this optimal target. Then the expected SSE payoff for the defender when facing this adversary of type γ with probability p^γ is $\hat{U}_d^\gamma(C)$ $= U_d^\gamma(t^*, C) \times p^\gamma$, and for the attacker $\hat{U}_a^\gamma(C) = U_a^\gamma(t^*, C)$.

3.3.3 Solution Method

We introduce the ERASER-C algorithm (Efficient Randomized Allocation of Security Resources with Constraints), which takes as input a security game in the compact form described in Sect. 3.3.2 and solves for an optimal coverage vector corresponding to a SSE strategy for the defender. We allow resources to be assigned to *schedules* covering multiple targets. The set of legal schedules $S = \{s_1, \ldots, s_l\}$ is a subset of the power set of the targets, with restrictions on this set representing scheduling constraints. We define the relationship between targets and schedules with the function $H : S \times T \rightarrow \{0, 1\}$, which evaluates to 1 if and only if t is covered in s. The defender's strategy is now an assignment of resources to schedules, rather than targets. Another important notion is the presence of *resource types*, $\Omega = \{\omega_1, \ldots, \omega_v\}$, each with the capability to cover a different subset of S. The number of available resources of each type is given by the function $\mathcal{R}(\omega)$. Coverage capabilities for each type are given by the function $Ca : S \times \Omega \rightarrow \{0, 1\}$, which is 1 if the type is able to cover the given schedule and 0 otherwise.[1]

The combination of schedules and resource types captures key elements of the security domains. For example, in FAMS, federal air marshals are resources, and flights are potential targets, with payoff values defined by risk analysis of the flight. Due to location and timing constraints, however, a marshal cannot be on all possible flights. For example, a marshal in New York cannot board flights flying out of Los Angeles. Legal schedules can be used to define the set of possible flights that a

[1] Our current implementation uses complete matrices to represent H and Ca, but sparse representations could offer additional performance improvements.

Table 3.2 Notation table

Symbol	Meaning
d^{γ}	Reward of defender against adversary of type γ
k^{γ}	Reward of adversary type γ
p^{γ}	Probability of occurrence of adversary of type γ
Γ	Set of adversary types
T	Set of targets
A^{γ}	Attack vector for the adversary of type γ
a_t^{γ}	Probability of adversary of type γ attacking target t
C	Coverage vector of the defender
c_t	Probability of defender covering target t
$h(s, \omega)$	Probability of coverage of schedule s by defender type ω
x_s	Total coverage probability over schedule s
S	Set of valid schedules
Ω	Set of resource types
$Ca(s, \omega)$	Capability: 1 if type ω can cover schedule s; 0 otherwise
$\mathcal{R}(\omega)$	Number of available resources of type ω
$H(s, t)$	Mapping: 1 if schedule s covers target t; 0 otherwise
M	Huge positive constant
$U_d^{\gamma}(t, C)$	Utility of the defender when facing adversary type γ who attacks target t when defender coverage is C
$U_a^{\gamma}(t, C)$	Utility of the adversary of type γ when target t is attacked and defender coverage is C

federal air marshal could fly, given these constraints. Resource types are use to define the initial state (notably, location) of a marshal, which defines a subset of legal schedules that any given marshal could fly.

Adding scheduling and resource coverage constraints reduces the space of feasible coverage vectors. Consider an example with a single federal air marshal defending three flights. Suppose that there are two legal schedules, covering targets {1, 2} and {2, 3}. Given only these schedules, it is not possible to implement a coverage vector that places 50% probability on both targets 1 and 3, with no coverage of target 2.

The algorithm is a mixed-integer linear program (MILP) described in (3.4)–(3.11), with notation presented in Table 3.2. Constraints (3.5) and (3.12) force each adversary to select a pure strategy attacking a single target. The coverage vector C is constrained by the number of available resources through (3.8) and the coverage in each target to be in the range [0, 1] by (3.13). The coverage of each schedule must sum to the contributions of the individual resource types, specified by constraint (3.6). The mapping between the coverage of schedules and coverage of targets is enforced in (3.7). Constraint (3.8) restricts the schedule so that only the available number of resources of each type are used. Constraint (3.9) enforces that no probability may be assigned infeasible schedules for each resource type. The defender's expected payoff is defined with constraint (3.10) when follower γ attacks target A^{γ}. Since the objective maximizes d^{γ}, for any optimal solution $d^{\gamma} = U_d^{\gamma}(C, A^{\gamma})$. This also implies that C is maximal, given A^{γ} for any optimal solution, since d^{γ} is maximized. In a similar way, constraint (3.11) forces the attacker to select a strategy in the attack set of C. If the attack vector specifies a target that is not maximal, this constraint is violated. Therefore, taken together, the objective and constraints (3.10)–(3.11) imply that C and A^{γ} are mutual best-responses for the defender and the adversary in any solution. Thus, the defender mixed strategy C and the adversary attack vector A^{γ} for each adversary type γ form an SSE of the security Stackelberg game.

$$\max_{a,c,q,h,d,k} \sum_{\gamma \in \Gamma} d^{\gamma} p^{\gamma} \tag{3.4}$$

$$\sum_{t \in T} a_t^{\gamma} = 1 \quad \gamma \in \Gamma \tag{3.5}$$

$$\sum_{\omega \in \Omega} h_{s,\omega} = x_s \quad s \in S \tag{3.6}$$

$$\sum_{s \in S} x_s H(s,t) = c_t \quad t \in T \tag{3.7}$$

$$\sum_{s \in S} h_{s,\omega} Ca(s,\omega) \leq \mathscr{R}(\omega) \quad \omega \in \Omega \tag{3.8}$$

$$h_{s,\omega} \leq Ca(s,\omega) \quad s,\omega \in S \times \Omega \tag{3.9}$$

$$d^{\gamma} - U_d^{\gamma}(t,C) \leq (1 - a_t^{\gamma}) \cdot M \quad t \in T, \gamma \in \Gamma \tag{3.10}$$

$$0 \leq k^{\gamma} - U_a^{\gamma}(t,C) \leq (1 - a_t^{\gamma}) \cdot M \quad t \in T, \gamma \in \Gamma \tag{3.11}$$

$$a_t^{\gamma} \in \{0,1\} \quad t \in T, \gamma \in \Gamma \tag{3.13}$$

$$c_t \in [0,1] \quad t \in T \tag{3.13}$$

$$x_s \in [0,1] \quad s \in S \tag{3.14}$$

$$h_{s,\omega} \in [0,1] \quad s,\omega \in S \times \Omega \tag{3.15}$$

The payoff values $U_{\mathrm{d}}^{\gamma}(t, C)$ and $U_{\mathrm{a}}^{\gamma}(t, C)$ are calculated based on (3.1) and (3.2). The values of $U_{\mathrm{d}}^{\gamma,c}$ and $U_{\mathrm{d}}^{\gamma,u}$ used in these equations are the payoff values to the defender when a target is covered and uncovered, respectively. These values are provided by the domain experts, as described in Sect. 3.5. Similarly, the payoff values for the adversaries are also provided by the domain experts.

The values of other model parameters are calculated based on the user input and the game specification. Police officers and canines are the resources for ARMOR for checkpoint and ARMOR for canine, respectively. ARMOR does not differentiate between different resources (e.g., all canines are assumed to be equally capable), and hence there is exactly one resource type Ω. The number of resources $\mathscr{R}$, i.e., checkpoints or canines, is directly input by the user in the system. In the case of ARMOR, the set of legal schedules is an assignment of a checkpoint to an inbound road and is automatically generated by the system since ARMOR is aware of the road map of the airport. The capability matrix Ca in ARMOR consists of all ones since any resource could be assigned to any target. For example, any canine could be scheduled to any terminal.

Similarly, all the model parameters are defined based on user input and domain constraints in IRIS. The federal air marshals are the resources for IRIS. In IRIS, the different FAMS Offices form the different resource types. This information has already been supplied to IRIS by the domain experts. The numbers of resources of each type $\mathscr{R}$, that is the number of federal air marshals in each office, is directly input in IRIS by the end users. The set of legal schedules S is provided as an input to the system by the FAMS in IRIS. Each schedule in IRIS is a sequence of flights that a federal air marshal can take to complete a tour. In IRIS, the capability matrix Ca is defined based on resource types; for example, federal air marshals at the FAM office based in Los Angeles can only cover schedules flying out of Los Angeles, and hence only those schedules would have their capabilities set to 1. The mapping M is also calculated by the systems based on the domain specifications. For example, in IRIS, if schedule s is to take flight f1 followed by flight f2, then the row in M corresponding to s would have ones only for columns corresponding to f1 and f2.

Kiekintveld et al. (2009) have shown that the ERASER-C MILP corresponds to an SSE of the security game. The intuition behind the proof are two claims: (1) the coverage probability of the leader and the attack set of the follower are mutual best-responses by the construction of the MILP, and (2) the coverage probability of the leader gives the leader the optimal utility.

3.4 Software Systems Deployed at the LAX and FAMS Domains

Both LAX and FAMS are security scenarios in which there is a leader/follower dynamic between the security forces and terrorist adversaries. In both domains there are limited resources available to protect a very large space of possible targets, so it is not possible to provide complete coverage. Finally, the targets have diverse values and vulnerabilities in each domain. The domains, however, differ primarily due to size. In the LAX security domain there are eight terminals that must be protected, while the air marshals are responsible for protecting tens of thousands of commercial flights each day. This difference in size requires, in addition to scalable solution algorithms, different types of interfaces to have domain experts specify each game. Finally, while in the LAX domain all security resources can reach all targets, in the FAMS domain, the security resources must satisfy more complicated constraints (e.g., a given marshal cannot be assigned to two flights with overlapping time schedules).

In this section, we describe both security domains (LAX and FAMS) and discuss the architecture of the software systems developed for these domains. We begin with a description of the generic software architecture and then describe each domain and their specific software assistant. We finish this section with a list of lessons learned in doing these deployments.

3.4.1 Software Assistants

We now describe in detail the system architecture for each of the two software assistants, focusing primarily on the ARMOR system but providing some discussion of IRIS as a point of comparison. We paid particular attention to organization acceptance during the development process. The end users of both ARMOR and IRIS are security officers, and the system must be simple enough for them to be comfortable using it on a regular basis. In particular, the systems are designed to hide as much of the complexity of the game-theoretic models as possible, while still allowing enough flexibility for the users to input important parameters that change regularly. This required considerable effort in both user interface design and identifying ways to simplify and reduce the inputs required by the system to specify a game model. In the case of IRIS, it was also very important to build in functionality to import data from other systems to ease the burden of data entry (e.g., importing flight information from existing databases). Finally, the schedules that the system produces must be presented in a format that is easy to understand, with tools that allow final modifications if necessary.

Both ARMOR and IRIS are stand-alone desktop applications. ARMOR was developed in the Microsoft.NET framework, while IRIS is a stand-alone Java application. Due to security concerns, both systems are run on machines that are not connected to any network. The underlying solution methods use the open source GLPK[2] toolkit to solve the necessary mixed-integer programs. The general structure of the two applications is shown in Fig. 3.1. The core architecture can be divided into three modules, which we describe in detail in the subsequent sections:

1. *Input*: Interface for the user to enter parameters and domain knowledge.
2. *Back-end*: Inputs are translated into a game model, which is passed to the Bayesian Stackelberg game solver and then to a final process that generates a specific sample schedule based on the computed probabilities.
3. *Display Module*: The final schedule is presented to the user, with options to modify the output if necessary.

---

[2] http://www.gnu.org/software/glpk/.

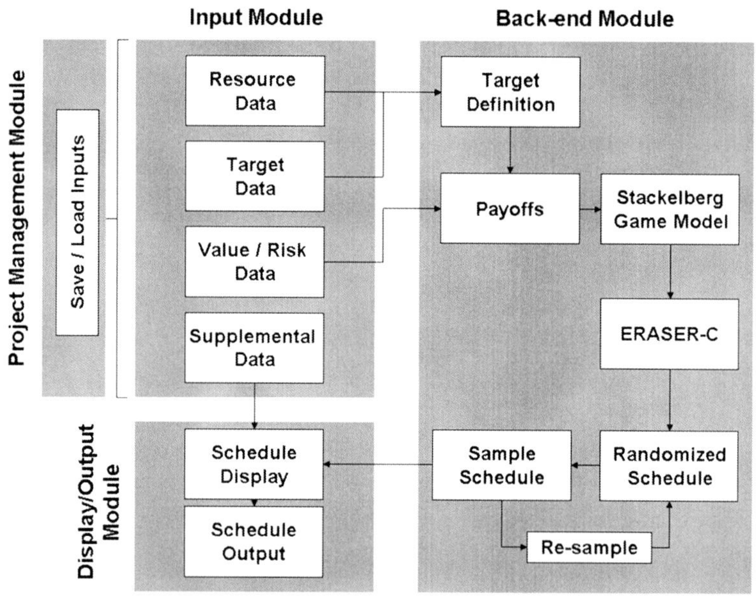

Fig. 3.1 General structure for the security assistants

We rely on the users and domain experts to provide the knowledge required to specify the game model. While some elements of the model do not change over time, others change frequently. For these, we must provide the users a convenient way to enter the necessary values. The basic inputs that both ARMOR and IRIS require fall into four categories: (1) the number of available resources and their capabilities, (2) the set of targets, (3) payoff values for each target, and (4) supplemental data to improve the user experience (e.g., names and labels). Both applications allow users to save and reuse this information across multiple executions.

The balance of how much information is hard-coded and how much is entered by the user is quite different for ARMOR and IRIS. For example, in ARMOR the set of targets is hard-coded because the number of terminals at LAX changes very rarely. However, in IRIS the flight information may change every time the system is run, so this is part of the user input. Determining which parameters were necessary to expose to the user was a significant task, and required several iterations with the domain experts and end users to strike the right balance between the complexity of the inputs and the flexibility of the system to capture the necessary information.

The Back-end module is fairly common to the two applications. This model builds a specific instance of a Bayesian Stackelberg game, based on all of the data provided by domain experts and entered through the GUI by end users. Some of the necessary information is hard-coded in each system, while other inputs can be modified by the user during the scheduling process.

Once an explicit game model has been generated, it is passed as input to the ERASER-C mixed-integer program. This model is solved using the standard open source solver GLPK in these applications. ERASER-C returns an optimal mixed strategy for the defender—a probability distribution over the defender's actions—which represents a randomized policy for allocating the security resources of either LAX or FAMS. We sample the randomized schedule found to generate a specific schedule for the security forces. This sample schedule specifies exactly where and when each resource should be assigned to each target. If necessary, it is also possible to "resample" from the randomized schedule to get another specific schedule, though this capability is used rarely. Any specific constraints that the schedules must satisfy are taken into consideration when final schedules are sampled. These sampled schedules are then displayed for the user through the Display Module.

The output module presents the generated sampled schedule to the user. The user can then review the schedule and accept it as is, or add additional constraints and run the scheduling process again. Since the specifics of Input and Display Modules are domain dependent we describe both of them, first for LAX and then for FAMS.

3.4.2 LAX Domain: ARMOR

LAX is the fifth busiest airport in the USA, the largest destination airport in the USA, and serves 60–70 million passengers per year (General description, 2007; Stevens et al., 2006). LAX is known to be a prime terrorist target on the west coast of the USA, with multiple arrests of plotters attempting to attack LAX (Stevens et al., 2006). To protect LAX, the airport police have designed a security system that utilizes multiple rings of protection. As is evident to anyone traveling through the airport, these rings include such things as vehicular checkpoints, police units patrolling the roads to the terminals, patrolling inside the terminals (with canines), and security screening and bag checks for passengers. Airport police use intelligent randomization within two of these rings: (1) placing vehicle checkpoints on inbound roads that service the LAX terminals, including both location and timing (a checkpoint is shown in Fig. 3.2), and (2) scheduling patrols for bomb-sniffing canine units at the different LAX terminals (as shown in Fig. 3.2). The numbers of available vehicle checkpoints and canine units are limited by resource constraints, so randomization is used as a method to increase the effectiveness of these resources while avoiding creating patterns in deployment.

The eight different terminals at LAX have very different characteristics, leading to different assessments of the value/risk for each terminal. For example, international flights are concentrated at a few terminals, while terminals have varying physical size and passenger loads. Because uncertainty about the adversary was identified by airport police as a key problem, the model should take into account the different types of adversaries that may be encountered. For example, there may be both hard-line, well-funded international terrorists planning attacks as well as

Fig. 3.2 Security checkpoints and canine patrols at LAX

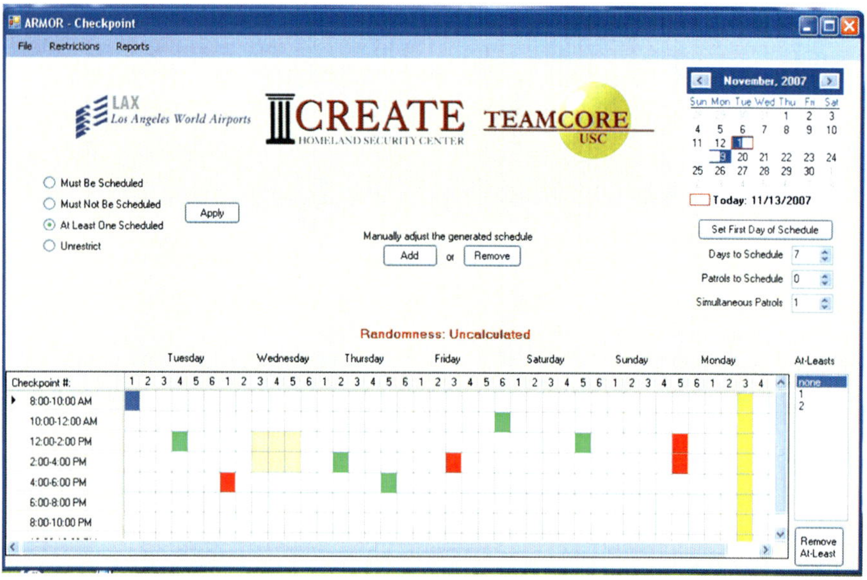

Fig. 3.3 ARMOR interface

amateur individuals. The payoff values for different attack scenarios should depend on the type of attacker and their capabilities.

The interface for the ARMOR checkpoints program is shown in Fig. 3.3 and provides options for the number of available resources, the number of scheduled days, the time slots to schedule, and the monthly calendar. A spreadsheet is used to display the proposed schedule and provide additional opportunities for the end users to modify the schedules in an iterative process. Three options are provided to change the possible scheduling actions: (a) number of checkpoints allowed during a particular time slot; (b) the time interval of each time slot; (c) the number of days to schedule over. Furthermore, three options are given to the user to enforce constraints onto the schedule: (a) forced checkpoint; (b) forbidden checkpoint; (c) at least one checkpoint. These constraints are intended to be used sparingly to accommodate situations where a user, faced with exceptional circumstances and extra knowledge, wishes to influence the output of the game. The user can impose these specific actions in the schedule using the spreadsheet interface. Each restriction is represented by a different color in the spreadsheet. The interface for the ARMOR Canine Patrols at LAX has similar features.

ARMOR generates a different game for each time slot on each day. The number of defender resources in the model is the number of canine units/checkpoints specified by the user. The number of targets is the number of terminals for the canines system, and the number of inbound roads for the checkpoints system. Generating the game matrix also requires values for the payoffs associated with each possible target. These payoff values depend on a variety of conditions, such as passenger loads, cost of the infrastructure, and publicity to the adversary. Domain experts provided us with formulae to automatically generate payoff values for all possible combinations of such conditions, which we encode in ARMOR. The system is also provided with estimates of the passenger load and other elements (the details of these formulae and tools cannot be discussed due to security concerns). For any given day, ARMOR is able to take the conditions for this day and select appropriate payoff values for the targets. As a result, it is not necessary for LAX police officers to enter these values by hand to generate each schedule, which is both time-consuming and error-prone. The system still retains a high degree of flexibility because values are precomputed and stored for a wide range of possible conditions.

The generated schedule of checkpoints and canines is presented to the user via a spreadsheet. Each row in the output spreadsheet corresponds to 1 h. Each column in the sheet corresponds to a terminal. Each entry in the sheet represents a schedule generated by ARMOR. The familiarity of the police officers with spreadsheets helped in the acceptance of the ARMOR schedules.

When ARMOR identifies that user constraints are causing unreasonably low likelihood of scheduling a checkpoint, it presents the schedule to the user with alerts. The user may then alter the schedule by modifying the forbidden/required checkpoints, or possibly by directly altering the schedule. Both possibilities are accommodated in ARMOR. If the user simply adds or removes constraints, ARMOR can create a new schedule. Once the schedule is finalized, it can be saved for actual use, thus completing the system cycle. This full process was designed to specifically meet the requirements at LAX for checkpoint and canine allocation.

3.4.3 FAMS Domain: IRIS

The FAMS places undercover law enforcement personnel aboard flight soriginating in and departing from the USA to dissuade potential aggressors and prevent an attack should one occur (TSA, 2008). The exact methods used to evaluate the risks posed by individual flights is not made public by the service, but we can identify many factors that might influence such an evaluation. For example, flights have different numbers of passengers, and some fly over densely populated areas while others do not. International flights also serve different countries, which may pose different risks. Special events can also change the risks for particular flights at certain times (Federal Air Marshal Service, 2008).

The scale of the domain is massive. There are currently tens of thousands of commercial flights scheduled each day, and public estimates state that there are thousands of air marshals. Air marshals must be scheduled on tours of flights that obey various constraints (e.g., the time required to board, fly, and disembark). Simply finding schedules for the marshals that meet all of these constraints is a computational challenge. Our task is made more difficult by the need to find a randomized policy that meets these scheduling constraints, while also accounting for the different values of each flight.

The FAMS domain is considerably larger, and the information required to build a game model in this domain changes much more frequently. For these reasons the application is considerably more complex than ARMOR in terms of the user interface and the mechanisms required to input all of the necessary information. This additional complexity is necessary in this domain to accurately capture the situation and provide all of the functionality requested by the end users. However, it does place a greater burden on the users to learn the system, and scheduling is a more time-consuming process than in ARMOR. Again, finding the right level of complexity to expose to the users was an iterative process that involved many discussions with the users and domain experts.

In the FAMS domain, we require information about the available air marshals, their scheduling constraints, the possible flights, and information about the risks/values to associate with each flight. The data about resources include information about the number and location of air marshals, as well as the conditions that define legal flight schedules. Flight information includes various data about each flight, including flight number, carrier, origin, destination, aircraft type, etc. Finally, some information is collected to improve usability, even though it is not strictly necessary for the game-theoretic analysis. This includes naming schemes for airports and airlines and other information that allows the system to output schedules in a more usable format or to interface easily with other systems. IRIS also includes functionality to import data from existing databases with flight data and other information. This greatly reduces the amount of data entry necessary to create a schedule.

Specifying the payoff values for every possible flight was a particular challenge in this domain, since there are thousands of flights to consider. We use an attributed-based system to elicit these values, based on the Threat, Vulnerability, and Consequence (TVC) model for estimating terrorism risk (Willis et al., 2005). By eliciting values for attributes of flights rather than specific flights, we are able to dramatically reduce the number of entries required by the user. Each flight is then given an aggregate value based on these components; the specific calculations used to determine flight risk are sensitive information and cannot be revealed. The values of the attributes for each flight can be populated automatically from existing databases. To allow for specific intelligence or exceptional circumstances, the individual payoff values for any flight can also be directly edited by the end user. However, this is only rarely necessary and the majority of the analysis can be effectively automated.

This preference elicitation system of IRIS has substantially reduced the number of values that must be entered by the user. During a restricted test run on real data, the attribute-based approach called for a total of 114 values to input regardless of the number of flights. By contrast, there were 2,571 valid flights over a week, each requiring four payoff values, summing to 10,284 user-entered values without the attribute-based preference elicitation system. The attribute-based approach clearly requires far fewer inputs and remains constant as the number of flights increases, allowing for excellent scalability as we deal with larger and larger sets of flights. Equally importantly, attribute-based risk assessment is an intuitive and highly scalable method that can be used in any problem where people must distill numerous attributes of a situation into a single value for a large number of situations that share the same attributes.

The generated schedules are presented to the user via the application window. The schedule created is shown in the interface, and allows the users to view more detailed information about each target. The user is also able to output the schedule to a file which can then be used to analyze the schedule in more detail. The sample assignment of federal air marshals to flight schedules is exactly a schedule that could be used by the FAMS. At this point, the scheduling assistant allows the expert using the system to create numerous sample schedules based on the same optimal mixed strategy or to change the assignment of federal air marshals to flight schedules by hand to create a final schedule that meets the needs of the FAMS. Of course, the user can also adjust any of the parameters entered and resolve the game completely. The output of IRIS is in the same format as the other systems used by the FAMS officers. It has not been presented here for simplicity and because of security concerns.

3.4.4 Lessons Learned

The design and deployment of ARMOR and IRIS have posed numerous challenges. We outline some key lessons learned during the design and deployment of these tools. First, there is a critical need for randomization in security operations. Security officials are aware that requiring humans to generate randomized schedules is unsatisfactory because, as psychological studies have often shown (Wagenaar, 1972; Treisman and Faulkner, 1987), humans have difficulty in randomizing, and they can also fall into predictable patterns. Instead, game-theoretic randomization that appropriately weighs the costs and benefits of different actions and randomizes with appropriate weights leads to improved results. Security officials were therefore extremely enthusiastic in their reception of our research and eager to apply it to their practices.

Second, organizational acceptance is a key issue. In creating solutions for people, we must be cognizant of how difficult it will be for a user to adopt our solution. Each deviation from existing methodology is a step away from the familiar that we must convince the user to accept. Instead of asking people to make numerous and sometimes unnecessary changes, minimizing these differences and complexities can help pave the way toward a successful implementation. For example, tweaking the GUI to achieve a look and feel that the user is familiar and comfortable with can help the user understand the system faster and better. Similarly, because infrastructural changes are often costly and/or time-consuming, ease of incorporating our work into their daily routine is essential. For example, using inputs and creating outputs that were in the same format as existing protocols minimized the additional work that our assistant would create for the security officers and lead to easier acceptance of the system.

Third, it is important to provide the users with operational flexibility. When initially generating schedules for canine patrols, we created a very detailed schedule, micro-managing the patrols. This did not get as positive a reception from the officers. Instead, an abstract schedule that afforded the officers some flexibility to respond to situations on the ground was better received.

3.5 A Generic Network Security Problem

As noted above, implementing a Stackelberg security game model to plan patrols is a difficult process that to date has been undertaken with substantial effort in collaboration with the security providers. In many situations, however, there is enough information about the security process that a data-driven process could be used to assist security providers in defining the actions and payoffs of the security game. In this section we illustrate recent work that aims to automatically build a Stackelberg security game for the problem of patrolling a street network to prevent crime. The proposed approach uses data-mining tools on a database of past reported crime and events to identify the locations to be patrolled, the times at which the game changes, and the types of adversaries faced. The idea is to exploit temporal and spatial patterns of crime on the area to be patrolled to determine the priorities on how to use the limited security resources.

We consider the street network depicted in Fig. 3.4 which corresponds to a centric commercial, turistic, and economic district in Santiago, Chile. This is a busy part of the city usually with large crowds on the street and that historically concentrates a high number of crimes, for the most part theft or minor aggressions. This type of crime in particular can be deterred or reduced with appropriate patrolling by police. To represent the problem of deciding where to patrol as a Stackelberg security game, security providers need to identify the specific points on this street network that concentrate crime and determine the payoffs defenders and attackers would receive if crimes at these locations are committed or are prevented. In this security game, police patrols on foot would go to the points selected

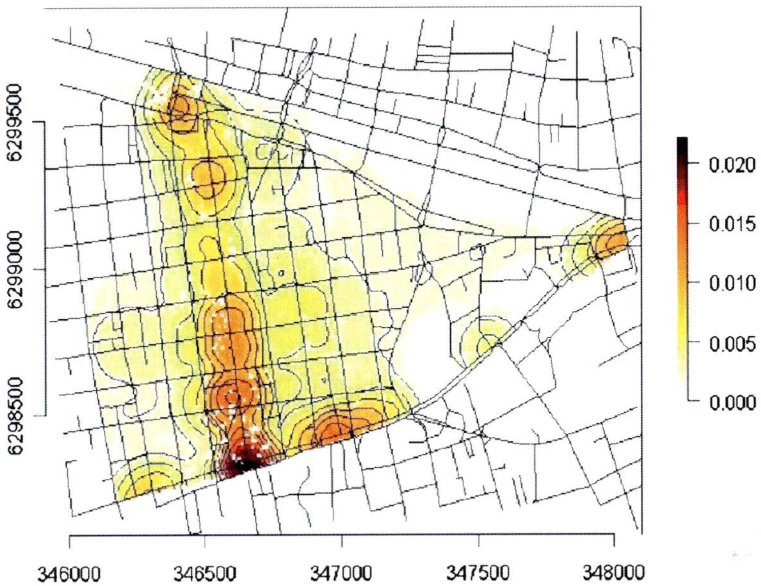

Fig. 3.4 Patrolling area with a density plot of reported crimes in the period 12/15/2002–12/14/2004

following the random optimal mixed strategy that maximizes the defender's utility. Different types of criminals would, knowing the optimal mixed strategy of the police patrols, then decide where to attack on the network, if at all. We assume that both police and criminals appear at the point selected, without interacting in other parts of the network. In addition to the description of the street network, we obtained from the Chilean national police force information about reported crimes in the area and the police reports for a 2-year period. Each reported crime has a location, a date and time, and a description of the crime (classification of crime [robbery, theft, etc.], amount stolen, level of violence, etc.). The police reports include information about the available resources in each shift, which helps estimate the police resources used for preventive patrolling.

3.5.1 Building a Data-Driven Security Game

This information is then processed in an automated data-driven procedure to build a security game in five steps: (1) Define the amount of data that will be relevant to calibrate the security game, (2) Determine locations to patrol, (3) Identify attacker types from data, (4) Determine times to patrol, (5) Determine payoffs for leader and followers.

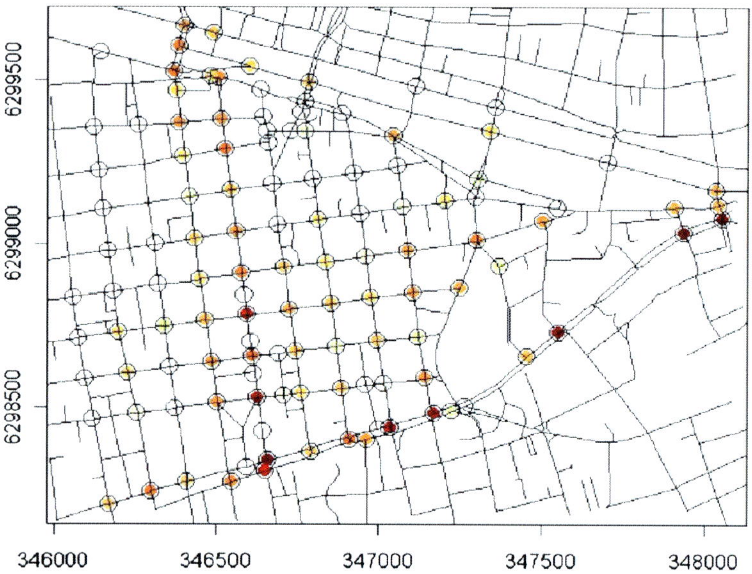

Fig. 3.5 Optimal mixed strategy on locations selected. Node color corresponds to probability of coverage of the node, with a *darker color* indicating a higher probability of coverage

Step 1: In determining which data to use to build a representative security game, we must strike a balance between selecting too much data and too little. Sufficient past data should be included so that significant but perhaps rare patterns of crime are taken into consideration. However, if too much data are taken into account, we run the risk of representing crime patterns that no longer exist. You must rely on expert opinion to estimate how representative past data are of the current security situation leading to an estimate of how much of the past information to use in identifying the locations of the patrols, the types of adversaries, and the utilities for each. In the results we show below, we used a time window of 2 years of data (from December 15, 2002 until December 14, 2004) to build a week long game (for the week of December 15–22, 2004).

Step 2: We used an off-the-shelf clustering software to identify the locations to be patrolled from the density plot of reported crime displayed in Fig. 3.5. These locations are selected anywhere on the road network in a way that summarizes the geographical distribution of crimes without requiring a massive number of locations. We used the software DBSCAN (density-based spatial clustering of applications with noise) (Ester et al., 1996). This is a density segmentation tool which also removes the noise in the data and automatically selects the number of segments to consider. In the results we obtained, DBSCAN identifies 119 locations to protect in which there are at least 10 crimes within a radius of 20 m. These points represent 89.23% of the reported crimes. We note that a number of good clustering algorithms can help in identifying a set of locations that are representative of the spatial crime distribution.

Step 3: We follow the knowledge discovery in databases (KDD) scheme (Fayyad et al., 1996) to process the database of reported crimes and identify different types of attackers. The KDD approach is a generic scheme that outlines a series of procedures to, among other things, create a target data set, remove data noise and outliers, handle missing data, identify useful features in the data, etc. Each of the processes can be implemented with any of a number of existing tools. For the selection of attributes, we chose a wrapper technique that automatically selects the attributes that help segmentation (Dy and Brodley, 2004). To identify the clusters of crimes we use a k-means clustering model. We found that this model was superior to alternative clustering models we tried (X-means, expectation maximization) for this problem, both in runtime and the quality of solutions found, which are more easily interpretable. The number of reported crimes in each cluster informs us of the frequency of different types of crime and thus the likelihood of facing each. The crimes in the 2-year database were classified into 9 significant clusters that were characterized by 24 significant attributes.

Step 4: Since the security conditions change during the day and the Stackelberg security game describes static conditions, we separate the day into different time intervals (or blocks) in which the security conditions remain almost constant. The different types of crime identified in Step 3 include three different time blocks which are found to be significant. Intersecting these times with the police patrolling shifts gives us a total of seven time intervals, or blocks, where the likelihood and composition of different types of crime and patrolling resources are kept about constant.

Block	From	To	Block	From	To	Block	From	To
S1	0:00	6:59	S2	7:00	9:59	S3	10:00	14:59
S4	15:00	17:59	S5	18:00	19:59	S6	20:00	21:59
S7	22:00	23:59						

In blocks S2 and S3 there are 23 patrolling units available, in blocks S4, S5, and S6 there are 24 patrolling units, and in blocks S1 and S7 there are nine patrolling units. Here, one patrolling unit corresponds to a pair of policemen on foot.

We determine the probability of facing each type of adversary by the frequency with which each of the nine types of crimes occur. To make this frequency more dependent on recent events, the past event data are scaled with an exponential decay function. Table 3.3 shows these frequencies for each of the nine types of crimes over the seven time blocks found.

Table 3.3 Probability of facing each follower in the different time blocks

Cluster	S1	S2	S3	S4	S5	S6	S7	Total
0	0.234	0.516	0.624	0	0.603	0.562	0.395	1,815
1	0.078	0.057	0.048	0.142	0.049	0.079	0.097	679
2	0	0	0	0.470	0	0	0	545
3	0.032	0.018	0.018	0	0.012	0.027	0.050	369
4	0	0	0	0.260	0	0	0	405
5	0.253	0.091	0.063	0.079	0.066	0.093	0.150	808
6	0.023	0.027	0.022	0.048	0.033	0.016	0.024	419
7	0	0	0	0	0	0.223	0.285	575
8	0.381	0.291	0.225	0	0.238	0	0	1,110
Total	727	457	1,892	1,217	939	881	612	

Table 3.4 Expected payoff for each type of criminal, in US dollars, if attack is successful (average utility) and if attack is unsuccessful (average cost, using a 40 % discount rate while in prison)

Cluster	Average utility	Prison time	Average cost
0	182	61	638
1	209	1,752	731
2	136	63	476
3	451	1,746	1,579
4	175	1,747	614
5	217	1,686	761
6	139	74	486
7	138	1,757	485
8	218	1,739	764

Step 5: In this work we determine the payoffs for the attacker as a valuation of the monetary payoff of being successful or getting caught for each type of crime. In the case of the police, we estimate that the payoff for catching a criminal is zero (for all types) while the penalty for a successful crime equals the expected amount earned by that type of criminal. We first determine from the information on the database the average expected reward for the criminal in a successful attack. To determine the penalty of an unsuccessful attack, we estimated the expected number of days in jail for that type of crime and evaluated the amount of forgone earnings for the criminal for not being able to commit crimes during that period. We note that there are a number of alternative models that can be incorporated here, in particular models of risk aversion that better represent human behavior in adversarial environments, such as prospect theory (Kahneman and Tvesky, 1979) or quantal response (McKelvey and Palfrey, 1995). Table 3.4 presents the values of payoffs for the Stackelberg game for each of the nine types of adversaries.

Table 3.5 Defender's expected utility in different time blocks

Block of time	S1	S2	S3	S4	S5	S6	S7
Stackelberg (mixed)	– 1.87	– 0.31	– 0.29	– 0.24	– 0.21	– 0.22	– 1.63
Stackelberg (pure)	– 8.87	– 3.90	– 3.62	– 3.27	– 3.53	– 3.48	– 8.30
Maximin (mixed)	– 5.40	– 2.42	– 2.21	– 1.94	– 2.18	– 2.18	– 5.10
Maximin (pure)	– 8.87	– 3.95	– 3.67	– 3.27	– 3.53	– 3.48	– 8.30

3.5.2 Additional Considerations in a Data-Driven Security Game

The procedure above helps security providers build a Stackelberg security game to determine efficient patrols in an urban street network. This game can then be formulated as the mixed integer programs described in Sect. 3.3 and solved to optimality. A solution for this problem is depicted in Fig. 3.5. The color at each node corresponds to the amount of coverage in the optimal mixed strategy for a certain time block. To implement this solution, the police should sample from this distribution to decide which locations to patrol each day in every time block.

The game developed can also be used to evaluate the current practice and the proposed patrol plan. Currently police direct their preventive patrols to the locations where the highest concentration of crime is expected to occur, based on recent past activity (2 weeks). We assume that the highest concentration of crime are the locations where the game predicts the highest payoff for the adversary, therefore directing the patrols to the maximum payoff locations leading to a minimax strategy. Table 3.5 presents the defender's expected profits in each time block under each of four different strategies: the optimal mixed and pure strategies of the Stackelberg game and Minimax. We note that the utility for the leader is always better in the Stackelberg game (mixed).

The set of tools described here hope to complement the experience and intuition of law enforcement. There is much information that is difficult to include in decisions on how to patrol. This is the case in part because of the amount of data and in part because the data are not being collected or are biased. We note that a better description of the security problem can be obtained, and thus a better security game formulated, by incorporating additional sources of information, such as surveys of victimization and physical description of places. We believe this is an interesting avenue of future research to create robust systems that would be more easily deployable in diverse settings.

3.6 Conclusions

Monitoring and patrolling are key components of law enforcement in security domains. In generating schedules for these patrols, it is important to account for varying weights of the targets being protected as well as the fact that potential attackers can often observe the procedures being used. This chapter describes scheduling assistants for the LAX police, ARMOR, and the FAMS, IRIS, which provide game-theoretic solutions to this problem. The two systems assist the security forces in generating randomized patrols while ensuring that differences in importance of different targets are preserved. A critical observation in the deployment of these scheduling assistants is the difficulty faced in reducing a complex security domain to a Stackelberg game model. To address this difficulty we present a data-mining-based model to assist security personnel in defining the Stackelberg security game from historic data.

ARMOR and IRIS make use of algorithmic advances in multi-agent systems research to solve the class of massive security games with complex constraints that were not previously solvable in realistic time-frames. Thus, although our applications were designed to be deployed at LAX and FAMS, they provide a general framework for solving patrolling scheduling problems in other domains as well.

Our approach of using Stackelberg games to model real-world security problems is applicable in a wide range of domains that share the following attributes: (a) there are intelligent players, (b) one player's strategy is observable by the other player, (c) player's have varying preferences among targets, and (d) it is not possible to provide full coverage of all targets. Some examples of similar security situations include security in computer networks, checkpoints at subway stations, security inspections at ports, and monitoring of other mediums of public transport.

Ultimately the security providers (Police, Air Marshals) are the judge of the usefulness of these Stackelberg security game models. As in any model it is critical to allow for expert knowledge to inform the system and provide feedback on the quality of solutions. With this in mind the development of the interface of these deployed systems has been an important aspect of this work. This research and these applications have been effective in helping in the security officers with scheduling and patrolling concerns. Thus, ARMOR and IRIS represent successful transitions of game-theoretic advances to applications that have been in use and effective in the real world. There are a number of additional improvements to these systems that could be done in the future to facilitate deployment to different domains. Some lines of future research include methods to incorporate qualitative information (estimates of unreported crime, fear of crime, etc.) to construct the Stackelberg games; coordination of different security resources; and considering attackers who deviate from rational behavior (due to differences in information or human bias).

References

Agmon N, Sadov V, Kaminka GA, Kraus S (2008) The impact of adversarial knowledge on adversarial planning in perimeter patrol. In: AAMAS

An B, Pita J, Shieh E, Tambe M, Kiekintveld C, Marecki J (2011) GUARDS and PROTECT: next generation applications of security games. ACM SIGecom Exchanges 10(1):31–34

Avenhaus R, von Stengel B, Zamir S (2002) Inspection games. In: Aumann RJ, Hart S (eds) Handbook of game theory, vol 3. North-Holland, Amsterdam, pp 1947–1987 (Chap. 51)

Babu L, Lin L, Batta R (2006) Passenger grouping under constant threat probability in an airport security system. Eur J Oper Res 168:633–644

Bard JF (1999) Practical bilevel optimization: algorithms and applications. Nonconvex optimization and its applications, vol 30. Springer, Berlin

Basar T, Olsder GJ (1995) Dynamic noncooperative game theory, 2nd edn. Academic, San Diego

Bier VM (2007) Choosing what to protect. Risk Anal 27(3):607–620

Blanco M, Valino A, Heijs J, Baumert T, Gomez JG (2007) The economic cost of March 11: measuring the direct economic cost of the terrorist attack on March 11, 2004 in Madrid. Terror Polit Viol 19(4):489–509

Breton M, Alg A, Haurie A (1988) Sequential stackelberg equilibria in two-person games. Optim Theor Appl 59(1):71–97

Brown G, Carlyle M, Kline J, Wood K (2005) A two-sided optimization for theater ballistic missile defense. Oper Res 53:263–275

Brown G, Carlyle M, Royset J, Wood K (2005) On the complexity of delaying an adversary's project. In: Golden B, Raghavan S, Wasil E (eds) The next wave in computing, optimization and decision technologies. Springer, Berlin, pp 3–17

Brown G, Carlyle M, Salmeron J, Wood K (2006) Defending critical infrastructure. Interfaces, 36 (6):530–544

Conitzer V, Sandholm T (2006) Computing the optimal strategy to commit to. In: ACM EC-06, pp 82–90

Dy JG, Brodley CE (2004) Feature selection for unsupervised learning. J Mach Learn Res 5:845–889

Ester M, Kriegel H-P, Sander J, Xu X (1996) A density-based algorithm for discovering clusters in large spatial database with noise. Technical report, Institute for Computer Science, University of Munich

Fayyad U, Piatetsky-Shapiro G, Smyth P (1996) From data mining to knowledge discovery: an overview. AI Mag 17(3):37–54

Federal Air Marshal Service (2008). http://en.wikipedia.org/wiki/Federal_Air_Marshal_Service

Fudenberg D, Tirole J (1991) Game theory. MIT, Cambridge

Gatti N (2008) Game theoretical insights in strategic patrolling: model and algorithm in normal-form. In: Ghallab M, Spyropoulos CD, Pakotakis N, Avouris N (eds) ECAI. IOS Press, Amsterdam, pp 403–407

General description: Just the facts (2007). http://www.lawa.org/lax/justTheFact.cfm

Harsanyi JC, Selten R (1972) A generalized Nash solution for two-person bargaining games with incomplete information. Manag Sci 18(5):80–106

Jain M, Tsai J, Pita J, Kiekintveld C, Rathi S, Ordóñez F, Tambe M (2010) Software assistants for patrol planning at LAX and federal air Marshals service. Interfaces 40(4):267–290

Jiang A, Leyton-Brown K (2006) A polynomial-time algorithm for action-graph games. Artif Intell 679–684

Kahneman D, Tvesky A (1979) Prospect theory: an analysis of decision under risk. Econometrica 47(2):263–292

Kiekintveld C, Jain M, Tsai J, Pita J, Tambe M, Ordóñez F (2009) Computing optimal randomized resource allocations for massive security games. In: Proceedings of the 8th International Conference on Autonomous Agents and Multiagent Systems. Budapest, Hungary, May 10–15, 1:689–696

Koller D, Milch B (2003) Multi-agent influence diagrams for representing and solving games. Games Econ Behav 45(1):181–221

Larson R (1974) A hypercube queueing modeling for facility location and redistricting in urban emergency services. J Comput Oper Res 1:67–95

Leitmann G (1978) On generalized stackelberg strategies. J Optim Theor Appl 26(4):637–643

Looney R (2002) Economic costs to the United States stemming from the 9/11 attacks. Strateg Insights 1(6)

Lye K-w, Wing JM (2005) Game strategies in network security. Int J Inf Secur 4(1–2):71–86

McKelvey RD, Palfrey TR (1995) Quantal response equilibria for normal form games. Games Econ Behav 10:6–38

Nie X, Batta R, Drury CG, Lin L (2007) Optimal placement of suicide bomber detectors. Mil Oper Res 12:65–78

Osbourne MJ, Rubinstein A (1994) A course in game theory. MIT, Cambridge

Paruchuri P, Tambe M, Ordonez F, Kraus S (2006) Security in multiagent systems by policy randomization. In: Proceedings of the Fifth International Joint Conference on Autonomous Agents and Multiagent Systems (AAMAS-06). Hakodate, Japan, May 8–12, 273–280

Paruchuri P, Pearce JP, Tambe M, Ordóñez F, Kraus S (2007) An efficient heuristic approach for security against multiple adversaries. In: Proceedings of the 6th International Joint Conference on Autonomous Agents and Multiagent Systems (AAMAS 2007). Honolulu, Hawaii, May 14–18

Paruchuri P, Pearce JP, Marecki J, Tambe M, Ordóñez F, Kraus S (2008) Playing games with security: an efficient exact algorithm for bayesian stackelberg games. In: Proceedings of the 7th International Conference on Autonomous Agents and Multiagent Systems. Estoril, Portugal, May 12–16

Pita J, Jain M, Western C, Portway C, Tambe M, Ordóñez F, Kraus S, Parachuri P (2008) Deployed ARMOR protection: The application of a game-theoretic model for security at the Los Angeles international airport. In: Proceedings of the 7th International Conference on Autonomous Agents and Multiagent Systems. Estoril, Portugal, May 12–16

Pita J, Tambe M, Kiekintveld C, Cullen S, Steigerwald E (2011) GUARDS – game theoretic security allocation on a national scale. In: Proceedings of the 10th International conference on autonomous agents and multiagent systems. Taipei, Taiwan, May 2–6, 1:37–44

Ruan S, Meirina C, Yu F, Pattipati KR, Popp RL (2005) Patrolling in a stochastic environment. In: 10th international command and control research and tech. symposium, June 13–16

Sandler T, Arce DG (2003) Terrorism and game theory. Simul Gaming 34(3):319–337

Srivastava V, Neel J, MacKenzie AB, Menon R, Dasilva LA, Hicks JE, Reed JH, Gilles RP (2005) Using game theory to analyze wireless ad hoc networks. IEEE Commun Surv Tutuor 7(4)

Stevens D, Hamilton T, Schaffer M, Dunham-Scott D, Medby JJ, Chan EW, Gibson J, Eisman M, Mesic R, Kelley CT, Kim J, LaTourrette T, Riley KJ (2006) Implementing security improvement options at Los Angeles international airport. http://www.rand.org/pubs/documented_briefings/2006/RAND_DB499-1.pdf

Thornton P (2005) London bombings: Economic cost of attacks estimated at 2bn. July http://www.independent.co.uk/news/business/news/economic-cost-of-attacks-estimated-at-1632bn-499281.html

Treisman M, Faulkner A (1987) Generation of random sequences by human subjects: Cognitive operations or psychological process? J Exp Psychol 116(4):337–355

TSA: Federal Air Marshals (2008). http://www.tsa.gov/lawenforcement/programs/fams.shtm.

Tsai J, Rathi S, Kiekintveld C, Ordóñez F, Tambe M (2009) IRIS - A tool for strategic security application in transportation networks. In: Proceedings of the 8th International Conference on Autonomous Agents and Multiagent Systems, Budapest, Hungary, May 10–15

von Stackelberg H (1934) Marktform und Gleichgewicht. Springer, Vienna

von Stengel B, Zamir S (2004) Leadership with commitment to mixed strategies. Technical report LSE-CDAM-2004-01, CDAM research report

Wagenaar WA (1972) Generation of random sequences by human subjects: A critical survey of literature. Psychol Bull 77(1):65–72

Wein LM (2009) Homeland security: From mathematical models to policy implementation: the 2008 Philip McCord Morse lecture. Oper Res 57(4):801–811

Willis H, Morral A, Kelly T, Medby J (2005) Estimating terrorism risk. RAND Corporation. Santa Monica. http://www.rand.org/pubs/monographs/2005/RAND MG388.pdf

Chapter 4
Interdiction Models and Applications

Nedialko B. Dimitrov and David P. Morton

Abstract Through interdiction models, we infer the vulnerabilities inherent in an operational system. This chapter presents four applications of interdiction modeling: (a) to delay an adversary's development of a first nuclear weapon; (b) to understand vulnerabilities in an electric power system; (c) to locate sensors in a municipal water network; and (d) to secure a border against a nuclear smuggler. In each case, we detail and interpret the mathematical model and characterize insights gained from solving instances of the model. We point to special structures that sometimes arise in interdiction models and the associated implications for analyses. From these examples, themes emerge on how one should model, and defend against, an intelligent adversary.

This chapter describes how to assess the vulnerabilities of operational systems by using interdiction models. We do so in the context of four applications from the literature: delaying an adversary's development of a first nuclear weapon; understanding vulnerabilities in an electric power system; locating sensors to rapidly detect an illicit contaminant injected in a municipal water system; and locating radiation sensors to detect a nuclear smuggler. The key steps in this approach involve answering the following questions: (1) How is the system operated? and (2) What are the vulnerabilities of that system? Operations research has a rich history of developing mathematical models to answer question (1). Key to our approach is that we must be able to answer question (1) when any subset of the system's components has been interdicted.

N.B. Dimitrov
Operations Research Department, Naval Postgraduate School, 1411 Cunningham Road, Monterey, CA 93943, USA

D.P. Morton (✉)
Graduate Program in Operations Research, The University of Texas at Austin, 204 E. Dean Keeton St, Stop C2200 Austin, TX 78712-1591, USA

J.W. Herrmann (ed.), *Handbook of Operations Research for Homeland Security*, International Series in Operations Research & Management Science 183, DOI 10.1007/978-1-4614-5278-2_4,

We may be the operators of the system of interest, or an adversary may operate the system or be operating within a system we own. The former case arises when the system involves critical infrastructure, such as an electric power system or municipal water system. In such situations, a third question arises: (3) How can we invest to make the system more resilient? The latter case arises, for example, when an adversary is managing a project or attempting to transport illicit material across our transportation network.

In answering question (1), we assume operation of the system optimally adapts after interdiction of a subset of system components. Here, interpret the term *interdiction* liberally. It can mean an action that removes or degrades one more more system components, e.g., damaging a generator, substation, or transmission line in an electric power system, or it can mean an action that delays completion of a task in a project. Interdiction can also mean detecting illicit operations on, or threats to, a system that we own. Interdiction models identify a set of system components to interdict, subject to resource limits, so that system performance is optimally degraded. The system components identified indicate the vulnerabilities of the system, answering question (2). Again, central to the analysis is the recognition that system operation will optimally adapt, post-interdiction, to the residual system.

The sections that follow develop four applications of interdiction modeling. We motivate each application, describe a mathematical model, discuss important modeling choices, discuss computational tractability, and describe insights from analysis utilizing the model. In Sect. 4.1, we discuss delaying the development of a nuclear weapon; in Sect. 4.2, we discuss identifying vulnerabilities in the nation's power grid; in Sect. 4.3, we discuss detecting deliberate contamination of our drinking water supply; and in Sect. 4.4, we discuss securing our nation's borders against illicit smuggling of nuclear material.

4.1 Delaying an Adversary's Nuclear Weapons Project

Preventing a nation from covertly developing a first nuclear weapon is an international priority. More countries are pursuing civilian nuclear energy, and that growth may continue given concerns both with volatility in fossil fuel supplies and global warming. There is apprehension that, with more states having civilian nuclear power programs, some states might pursue a clandestine enrichment and reprocessing program for the purpose of developing a nuclear weapon (McGoldrick, 2011). Once the international community detects an illicit program, the tools available to stop or delay proliferation include diplomatic actions, economic embargoes, embargoes of key technologies, poaching of key personnel, sabotage, and military strikes. How should the potential effectiveness of such options be evaluated?

In a pair of papers, Harney et al. (2006) first build a detailed operational model of how a "proliferator" would manage the complex project of building a first nuclear weapon, or rather, a small batch of such weapons. Then, Brown et al. (2009) formulate a model on top of that operational model in which an "interdictor" selects

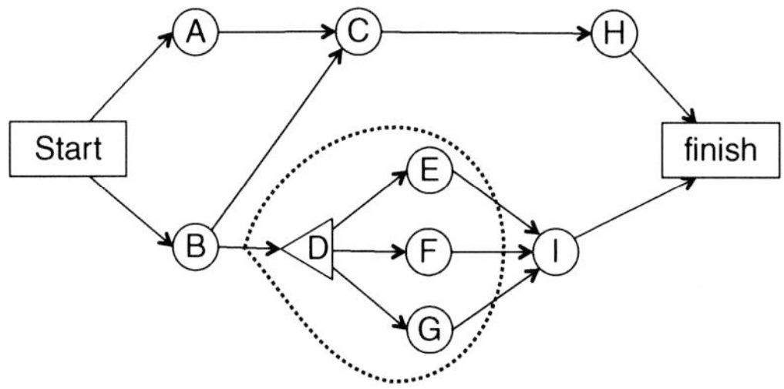

Fig. 4.1 An example PERT network. The nodes, A through I, denote tasks that have a duration. *Arcs* denote precedence relationships. For example, to start task C both tasks A and B must finish. The *triangular* node, D, and its successors, E through G, denote a decision task. Only one of the three tasks E, F, or G must be completed to begin task I, and in order to finish the project

a resource-constrained set of tasks to interdict so as to maximally delay completion of the nuclear weapons project. Reed (1994) describes a lower-fidelity model but one with a similar notion of interdiction. We summarize this line of work in this section.

4.1.1 Project Management for a First Nuclear Weapon

The natural modeling framework for representing how the proliferator would manage the project of building a first nuclear weapon is that of the program evaluation review technique (PERT)/critical path method (CPM). O'Brien (1969) discusses the origins of CPM and PERT, with the former beginning at Du Pont in 1956, and the US Navy developing the latter in 1958 in conjunction with the Polaris missile program. These tools have since been employed pervasively in industry and government (Shtub et al., 2005).

In its simplest form a PERT network models a collection of tasks represented by nodes that have specified durations and precedence relationships represented by directed arcs that indicate prerequisites. The length of the longest path in this network, called the critical path, indicates the minimum time required to complete the project. This time is achieved if all tasks on the critical path experience no delay, start as soon as possible, and tasks that are off the critical path are not sufficiently delayed.

Over the last several decades, PERT/CPM methods have evolved to capture the features needed to schedule and manage large, complex projects. Figure 4.1 depicts a simple PERT network with one such improvement: The "decision node" (node D in the figure) points to three alternative means to accomplishing a task. Brown et al. model three alternative technologies available to the proliferator to enrich uranium (gas centrifuges, gaseous diffusion, and aerodynamic enrichment) as well as these further enhancements:

1. In addition to the standard finish-to-start precedence relationship, start-to-start, finish-to-finish, and start-to-finish relationships are included. So in Fig. 4.1, task C might have a start-to-start precedence relationship with task B with the addition of a 10-week lead time, meaning the earliest task C can start is 10 weeks after the start of task B.
2. In addition to consuming time, tasks consume resources including energy, raw materials, and three types of labor: scientific, skilled, and unskilled. Consumption of these resources, in turn, consumes a monetary budget. We assume that the proliferator's *capability*, i.e., his level of each of these resources, is known. To understand the sensitivity of the results to the assumed capability, we run several analyses, varying the capability assumptions.
3. The proliferator can expedite, or "crash," tasks, subject to resource limits. A task has a nominal duration, $\bar{d}$, which the proliferator can decrease to $\underline{d}$. Crashing is assumed to be linear so that c_r units of resource r are consumed per week, say, that the task is expedited.

To simplify presentation of the proliferator's operational model, we choose to neglect some of the key fidelity introduced in Brown et al. (2009). We do not include start-to-start, finish-to-finish, and start-to-finish precedence relationships. We further neglect a combinatorial aspect of the proliferator's problem in which he must make decisions as to alternative means of accomplishing tasks, as indicated in the decision node D of Fig. 4.1. The simplified proliferator's operational model can be formulated as follows.

Sets	
$i, j \in N$	Nodes representing tasks
$(i, j) \in A$	Precedence relationships; task i must finish before j can start
$r \in R^{\mathrm{P}}$	Proliferator resources
Data [units]	
$\bar{d}_i$	Nominal duration of task i [weeks]
$\underline{d}_i$	Duration of task i if it is maximally expedited [weeks]
c_{ir}	Unit consumption of resource r for expediting task i [\$/week]
b_r	Budget of resource r for expediting tasks [\$]
lag_{ij}	Lag time between the completion of i and the start of j [weeks]
Proliferator's decision variables [units]	
S_i	Earliest start time of task i [weeks]
E_i	Time by which task i is expedited [weeks]

We distinguish two special nodes in N denoted "start" and "finish" as shown in Fig. 4.1. These artificial tasks have zero duration and consume no resources.

$$\min_{S,E} \quad S_{finish} \tag{4.1a}$$

$$\text{s.t.} \quad S_j - S_i \geq (\bar{d}_i - E_i) + lag_{ij}, \ (i,j) \in A \tag{4.1b}$$

$$\sum_{i \in N} c_{ir} E_i \leq b_r, \ r \in R^P \tag{4.1c}$$

$$0 \leq E_i \leq \bar{d}_i - \underline{d}_i, \ i \in N \tag{4.1d}$$

$$S_i \geq 0, \ i \in N \tag{4.1e}$$

$$S_{start} \equiv 0. \tag{4.1f}$$

Based on a finish-to-start precedence relationship, constraint (4.1b) indicates that the earliest start time for task j is the sum of: the earliest start time for task i, the duration of task i, and any additional lag time between the completion of i and the start of j. Constraint (4.1c) limits the consumption of resources, which have been allocated for expediting tasks. These resources include energy, raw materials, and different types of labor. Carrying out the tasks at their nominal durations also consumes resources, but that consumption has already been accounted in the b_r values. Constraint (4.1d) limits the magnitude by which each task can be expedited, and the time at which the project finishes is minimized in (4.1a).

4.1.2 Formulation and Tractability in Delaying a Project

Given model (4.1), under which the proliferator is presumed to operate, the interdictor seeks to maximally delay completion of the proliferator's project. The interdictor is the nation, or group of nations, determined to delay the proliferator. The interdictor is limited by a monetary budget and diplomatic constraints as well as limits on economic and environmental consequences. Logical constraints on interdiction are also easily incorporated. For example, perhaps at most one of the tasks A, C, and E can be interdicted. The interdictor's model requires the following additional constructs.

Sets	
$r' \in R^{\mathrm{I}}$	Resources for interdiction
Data [units]	
$v_{ir'}$	Consumption of resource r' from interdicting task i [\$]
$w_{r'}$	Budget of resource r' for interdicting tasks [\$]
delay_i	Delay in completing task i from its interdiction [weeks]
Interdictor's decision variables	
X_i	Binary variable that takes value 1 if task i is interdicted and 0 otherwise

The set of feasible interdiction plans is given by

$$\mathcal{X} = \left\{ X : \sum_{i \in N} v_{ir\prime} X_i \leq w_{r\prime}, r\prime \in R^I, X_i \in \{0,1\}, i \in N \right\}.$$

The set $\mathcal{X}$ can include further constraints such as the logical implications just mentioned. We let $f(X)$ denote the optimal value of model (4.1), except that the right-hand side of constraint (4.1b) is replaced by

$$(\bar{d}_i - E_i + delay_i X_i) + lag_{ij},$$

in which the duration to complete task i is modified to include delay$_i$ if task i is interdicted, i.e., if decision variable $X_i = 1$. The interdictor's optimization problem is then

$$\max_{X \in \mathcal{X}} f(X). \tag{4.2}$$

Subject to the constraints dictated by $\mathcal{X}$, the interdictor in model (4.2) seeks to maximally delay completion of the proliferator's project. The nested max–min structure of model (4.2), with $f(X)$ defined via the optimal value of model (4.1), modified to incorporate delays, means the interdictor:

1. First chooses one or more tasks to interdict.
2. After the interdictor's plans are revealed, the proliferator chooses a plan to best expedite his project and, in the full version of model (4.1) in Brown et al. (2009), the proliferator further chooses an enrichment technology associated with a decision node in the PERT network.

There are a number of questions that one can pose, regarding the appropriateness of the model just put forward. They include:

- Will the proliferator actually use PERT/CPM in planning his project?
- What if the proliferator behaves differently?
- Can the interdictor adapt his interdiction plan over time?
- Has the proliferator already committed to some decisions?

While the tools of PERT/CPM have been well known since the 1950s, and there are widely available commercial products that ease their application, we do not know whether the proliferator will employ these off-the-shelf tools. That said, the proliferator *is* building a nuclear weapon, and so it is arguably reasonable to assume he will use such tools to manage that project. If the proliferator behaves suboptimally, he will finish the project later than what we predict, and in this sense, our prediction of the induced delay is appropriately conservative. In reality, the interdictor could adapt his interdiction plan over time, and the proliferator may commit to a partial course of action prior to when some decisions must be made by the interdictor. The two-stage model we have sketched is more computationally tractable than a richer multistage model. The two-stage model is also conservative in the same sense that we have just mentioned: If the model predicts a completion time induced by a set of interdiction activities then the actual completion time is at least that long.

4.1.3 Practical Implications and Insights

Harney et al. (2006) and Brown et al. (2009) describe a PERT/CPM model instance with about 200 tasks and 600 precedence relationships for the proliferator's project. The model is similar to model (4.1) but with the enhancements discussed above. The proliferator has a budget of $380 million that can be allocated to the five types of resources: energy, raw materials, and scientific, skilled, and unskilled labor. Brown et al. build an interdiction model in the form of (4.2); they label the types of interdiction they consider as mild, nonmilitary delays; and, they introduce a limit on the total number of tasks that can be interdicted.

If the interdictor does nothing, then the proliferator completes the project in about 260 weeks to about 5 years. The same result holds if the proliferator's budget grows from $380 to $480 million.

Next, Brown et al. (2009) assume that the proliferator plans his project, ignoring the possibility of interdiction. Knowing this, the interdictor selects two tasks to interdict. Then, post-interdiction, the proliferator does not adapt his plans. In this case, the proliferator completes his project in 356 weeks, a delay of 37 %. This analysis runs counter to the notion of planning conservatively, which we sketch at the end of Sect. 4.1.2, but its value will become clear shortly.

Now suppose the interdictor selects two tasks to interdict, assuming the proliferator will adapt his plan optimally post-interdiction, and the proliferator does indeed respond optimally post-interdiction; i.e., we are in the setting of model (4.2). In this case, the proliferator completes his project in 348 weeks, a delay of 34 % over the nominal 260 weeks. The proliferator saves only 8 weeks by reacting optimally to the interdiction of two tasks, relative to not reacting at all. This may counter our intuition that optimal adaptation by the proliferator should better insulate his project from delay. As Brown et al. indicate, this means that the interdictor has uncovered "unavoidable fragilities" in the proliferator's project.

Further analysis shows that as the number of tasks to interdict ranges from one to four, the interdictor can delay the project from 1.1 to 2.4 years when the proliferator has a budget of $380 million. This range changes to 1.1–2.25 years when the proliferator has an additional $100 million. When examining the tasks selected for interdiction, we see that these tasks need not be on the (original) critical path. Finally, the specific task of "cascade loading" is a task that is interdicted in all the variants that the authors consider, a fact clearly of interest to decision makers.

Since the publication of the pair of papers (Brown et al., 2009; Harney et al., 2006) discussed above, there have been a number of related developments reported in the popular press. From January 2010 through January 2012, five Iranian nuclear scientists have been attacked, and four killed, most with a bomb magnetically attached to their cars by motorcyclists; see, e.g., Borger (2012). The so-called Stuxnet worm targeted control software for centrifuges used to enrich uranium, and it has been reported as being the most sophisticated malware ever developed (see, e.g., Broad et al. 2011; Keizer 2010). In September 2007, Israeli Air Force jets bombed what is reported to have been a partially developed nuclear reactor in Syria (see, e.g., Kessler and Wright 2007). According to the web site WikiLeaks, in a diplomatic cable dated April 20, 2008, between the US Embassy in Riyadh and Washington, the Saudi ambassador to the USA, Adel al-Jubeir, is reported as having "recalled the King's frequent exhortations to the US to attack Iran and so put an end to its nuclear weapons program." al-Jubeir is reported to have said, "He told you to cut off the head of the snake." WikiLeaks documents further indicate that then Secretary of Defense Robert Gates indicated that any such strike on Iran would not eliminate their nuclear program. Rather, it would only delay their pursuit of a nuclear weapon "by one to three years."

4.2 Vulnerabilities in the Electric Power Grid

In August 2003, a power surge blacked out parts of eight states in the northeast USA (Barron, 2003). The nation's essential demand for electricity and our dependence on the electric power grid to deliver electricity have been recognized in congressional assessments (Office of Technology Assessment U.S. Congress, 1990), and presidential policy planning groups (National Energy Policy Development Group, 2001). A congressional assessment dating back to 1990 states that "The bulk power system is vulnerable to terrorist attacks targeted on key facilities. Major metropolitan areas and even multi-state regions could lose virtually all power following simultaneous attacks on three to eight sites... " (Office of Technology Assessment U.S. Congress, 1990). In addition, the growth in demand for electric power is outpacing the development of new generation (National Energy Policy Development Group, 2001), while new sources of power such as wind and solar have highly variable generation and are difficult to integrate into the system (GE Energy for The National Renewable Energy Laboratory, 2010). All of these developments highlight the need to understand the vulnerabilities in our electric power grid.

At a high level, there are two approaches to analyzing the sustained operation of an electric power grid. The first approach studies the *reliability* of the power grid against random component failures (Rausand and Høyland, 2003). Such analysis uses data on the failure of individual components, such as wind turbines (Tavner et al., 2007), to analyze the system as a whole (Talukdar et al., 2003). However, it has been observed that the power grid can be quite robust to random failures, and at the same time be quite susceptible to the failure of a small number of select components (Albert et al., 2004). This leads to the second analysis approach, to study the *vulnerability* of the power grid against worst case component failure—rooted in the idea that an adversary could select components to attack to induce maximum disruption.

In the literature, there are two main methods of analyzing power-grid vulnerability. The first method defines some intuitive measures of grid component criticality, and then ranks the grid components based on those measures (Albert et al., 2004; Chassin and Posse, 2005; Espiritu et al., 2007; Qiang and Nagurney, 2008). The advantages of this method are that it does not require extensive computation, and the criticality measures can be adapted to, or borrowed from, other common networks. The disadvantage of this method is that it is not based on the physical properties of the electric power grid; that is, it is not based on the actual performance of the system. In particular, often these methods may not be validated with, or derived from, power-flow models for electricity distribution. Furthermore, as we discuss shortly with a specific example, the concept of ranking components based on criticality is flawed because criticality is a property of sets of power-grid components. A single component may not be critical on its own, but in a set of two components it may be highly critical.

The second approach is to use interdiction models for optimal or near-optimal interdiction of power-flow as in a pair of papers by Salmerón et al. (2004, 2009). The benefit of this approach is that it is indeed based on the physical properties of a given electric grid. An initial drawback of the approach is that it required algorithmic and computational tools that were not available when these researchers began studying the problem. Finally, to make the interdiction computations tractable, a steady-state optimal power-flow model is typically assumed. Specifically, the power-flow models used in interdiction do not include cascading failures. Steady-state power-flow models are good for modeling mid-term or long-term failures of the electric grid, on the range of weeks or months, as opposed to models of cascading failures (Mili et al., 2004), which cause outages on much shorter time scales.

For the remainder of this section, we outline the use of interdiction models to assess the vulnerabilities of the electric power grid (Salmerón et al., 2004, 2009). We focus on the intuitive interpretation of the models, basic mathematical formulation, and results from such analysis.

4.2.1 Interdicting the Electric Power Grid

To assess the vulnerability of an electric power grid, we seek to identify components that are critical to the continued operation of that grid. At a basic level, the electric power grid consists of transmission lines, buses, generators, and substations. The core generation and transmission components of the electric power grid connect to local power distribution networks, which operate at a lower voltage and deliver power to consumers. Some components of the grid, such as a transmission line, are relatively easy to repair. Others, such as generators or substations, could take weeks or months to repair, depending on the nature of the damage. Each component integrates with the grid in a unique fashion, based on the specific structure of the grid in question.

A natural way to measure the criticality of power-grid components is to understand how much disruption is caused by their loss. For example, the loss of a local power distribution line may cause a power outage to a dozen houses for a few hours, while the loss of a substation or generator may cause a power outage to a small city for weeks. We capture this intuitive measure of criticality through the *load shedding*, or the total demand for power that is not met if the component is lost and requires repair. We can define the criticality of a set of grid components in a similar fashion. The criticality of a set of components is the total load shedding if that set of components is lost and requires repair.

As is common in most interdiction models, the criticality of a set of grid components is not simply the sum of each component's criticality, but is instead determined by the structure of the grid in question and the corresponding system performance. For example, consider a town that is powered by two identical generators that each have enough capacity to satisfy the town's entire demand on their own. In addition, suppose that each generator is connected to the town's distribution network through its own transmission lines. The criticality of each generator on its own is rather low, because the other generator serves as a backup and there is no load shedding in the event of the loss of a single generator. However, the criticality of both generators as a set is very high, because if both generators are lost at once, all of the town's demand for power goes unserved.

Finding the set of k most critical components requires us to solve a complex combinatorial optimization problem. First, we require a model of how load shedding is affected by the loss of subsets of grid components. Second, we have to compute the set of k components that maximizes that load shedding. Salmerón et al. (2004, 2009) develop such a model and the associated algorithms for computing the most critical components.

4.2.2 Formulation and Tractability

The sequence of papers leading to the ability to analyze vulnerabilities of realistic electric power grids provides an excellent example of the development of interdiction models over time. The sequence begins with a simplified physics-based power-flow formulation (Salmerón et al., 2004), and builds to higher-fidelity power-flow models and the ability to analyze grids on the order of 5,000 buses, 5,000 transmission lines, and 1,000 transformers—the size of a large regional grid (Salmerón et al., 2009).

At the most basic level, a DC power-flow model is used to measure grid performance. Let $i \in I$ denote buses, $g \in G$ generators, $\ell \in L$ transmission lines, $c \in C$ consumer demand sectors, and $s \in S$ substations. Additionally, let $i \in I(s)$ denote buses at substation s, $g \in G(i)$ generation units connected to bus i, $\ell \in L_i^{\text{bus}}$ lines connected to bus i, $\ell \in L_s^{\text{sub}}$ lines connected to substation s, and $\ell \in L_{\ell'}^{\text{par}}$ lines running parallel to line ℓ'. In the model, transformers are represented by lines.

To fully specify the grid's structure, we also require a number of data parameters. Let $o(\ell)$ and $d(\ell)$ denote the origin and destination buses of line ℓ. Let $i(g)$ denote the bus for generator g, $\bar{P}_\ell^{line}$ be the transmission capacity for line ℓ, and $\bar{P}_g^{gen}$ be the maximum output of generator g. Let r_ℓ and x_ℓ be the resistance and reactance of line ℓ, giving a susceptance of $B_\ell = x_\ell / (r_\ell^2 + x_\ell^2)$. Finally, let d_{ic} be the demand for power (load) of consumer sector c at bus i and f_{ic} be the load-shedding cost for consumer sector c at bus i. We can also think of f_{ic} as specifying the relative importance of unmet demand in different consumer sectors, where shedding load at a hospital may be more costly than in another sector.

The DC power-flow model solves for the required generation from each generator (P_g^{gen}), the power-flow on each line (P_ℓ^{line}), the phase angle at bus i (θ_i), and the load shedding in consumer sector c at bus i (S_{ic}). Generating power in some generators is cheaper than others. It is also possible to introduce costs per generator and include those in the model, but we leave out that detail for simplicity. For brevity, let P denote the vector of variables P_g^{gen}, P_ℓ^{line}, θ denote the vector with components θ_i, and S denote that of S_{ic}. The DC power-flow can be computed using the following linear program:

$$\min_{P,\theta,S} \sum_{i\in I}\sum_{c\in C} f_{ic}S_{ic} \tag{4.3a}$$

$$\text{s.t.} \quad P_\ell^{\text{line}} = B_\ell(\theta_{o(\ell)} - \theta_{d(\ell)}), \quad \ell \in L$$

$$\sum_{g\in G(i)} P_g^{gen} - \sum_{\ell|o(\ell)=i} P_\ell^{line} + \sum_{\ell|d(\ell)=i} P_\ell^{line} \tag{4.3b}$$

$$= \sum_{c\in C}(d_{ic} - S_{ic}), \quad i \in I \tag{4.3c}$$

$$-\bar{P}_\ell^{line} \le P_\ell^{line} \le \bar{P}_\ell^{line}, \quad \ell \in L \tag{4.3d}$$

$$0 \le P_g^{gen} \le \bar{P}_g^{gen}, \quad g \in G \tag{4.3e}$$

$$0 \le S_{ic} \le d_{ic}, \quad i \in I, c \in C. \tag{4.3f}$$

The objective function of the linear program, (4.3a), minimizes load shedding. Constraint (4.3b) approximates active power flow on each line through a linear approximation involving the phase angles on the right-hand side; constraint (4.3c) maintains power balance at each bus; constraint (4.3d) maintains the transmission capacity of each line; constraint (4.3e) maintains the generating capacity of each generator; and, constraint (4.3f) enforces that load shedding cannot exceed demand.

Once the power-flow model is formulated, we place an interdiction model on top of that model to compute the k most critical components. For the interdiction model, we introduce binary variables that indicate whether each component is functional (binary variable is 0) or not (binary variable is 1). For the power-flow model (4.3), let the binary interdiction variables be δ_g^{gen}, δ_ℓ^{line}, δ_i^{bus}, and δ_s^{sub}, each indicating whether the corresponding system component is functional. Using the interdiction variables, we can compute if a transmission line ℓ is down and store the result in a binary variable d_ℓ (value of 0 for "no power," 1 for "may have power") as follows:

$$d_\ell = (1-\delta_\ell^{line})(1-\delta_{o(\ell)}^{bus})(1-\delta_{d(\ell)}^{bus}) \prod_{s|\ell\in L_s^{sub}} (1-\delta_s^{sub}) \prod_{\ell\prime|\ell\prime\in L_\ell^{par}} (1-\delta_{\ell\prime}^{line}). \quad (4.14)$$

Equation (4.4) states that transmission line ℓ cannot have power if line ℓ itself is nonfunctional; its origin or destination buses are nonfunctional; any substation that the line connects to is nonfunctional; or any parallel line is nonfunctional. The variable d_ℓ is only for notational convenience and is not an interdiction variable itself. For brevity, let the vector of interdiction variables for all components be δ. Using the interdiction variables and the notational convenience of d_ℓ, we can compute the k most critical components by solving the following optimization problem:

$$\max_{\delta,d} \quad \min_{P,\theta,S} \sum_{i\in I}\sum_{c\in C} f_{ic} S_{ic} s.t. \quad (4.5a)$$

$$\sum_{g\in G} \delta_g^{gen} + \sum_{\ell\in L} \delta_\ell^{line} + \sum_{i\in I} \delta_i^{bus} + \sum_{s\in S} \delta_s^{sub} = k$$

$$d_\ell = (1-\delta_\ell^{line})(1-\delta_{o(\ell)}^{bus})(1-\delta_{d(\ell)}^{bus}) \quad (4.5b)$$

$$\prod_{s|\ell\in L_s^{sub}} (1-\delta_s^{sub}) \prod_{\ell\prime|\ell\prime\in L_\ell^{par}} (1-\delta_{\ell\prime}^{line}), \quad \ell\in L \quad (4.5c)$$

$$\delta_g^{gen}, \delta_\ell^{line}, \delta_i^{bus}, \delta_s^{sub}, d_\ell \in \{0,1\}, \quad g\in G, \ell\in L, i\in I, s\in S \quad (4.5d)$$

$$P_\ell^{line} = B_\ell(\theta_{o(\ell)} - \theta_{d(\ell)}) d_\ell, \quad \ell\in L \quad (4.5e)$$

$$\sum_{g\in G(i)} P_g^{gen} - \sum_{\ell|o(\ell)=i} P_\ell^{line} + \sum_{\ell|d(\ell)=i} P_\ell^{line} = \sum_{c\in C}(d_{ic} - S_{ic}), \quad i\in I \quad (4.5f)$$

$$-\bar{P}_\ell^{line} d_\ell \le P_\ell^{line} \le \bar{P}_\ell^{line} d_\ell, \quad \ell\in L \quad (4.5g)$$

$$0 \le P_g^{gen} \le \bar{P}_g^{gen}(1-\delta_{i(g)}^{bus})(1-\delta_g^{gen}), \quad g\in G \quad (4.5h)$$

$$0 \le S_{ic} \le d_{ic}, \quad i\in I, c\in C. \quad (4.5i)$$

Like model (4.2) in Sect. 4.1, the interdiction model (4.5) is a bi-level program known as a Stackelberg game, with a nested "min–max." First, components are removed from the power grid. This is accomplished by binary decision variables δ subject to the cardinality constraint (4.5b) and binary restrictions (4.5d) along with the notational convenience, d, defined via constraint (4.5c). Second, using the remaining components, i.e., the residual power grid, variables P, θ, and S compute an optimal power flow to minimize load shedding. These variables are subject to constraints (4.5e)–(4.5i), which are similar to those of model (4.3), with additional parameterization in d. Specifically, the d_ℓ in constraints (4.5e) and (4.5g) ensure that no down transmission line can have power. Constraint (4.5h) similarly ensures that a disconnected or nonfunctional generator cannot generate power.

We can alter constraint (4.5b) to make interdicting some components more costly than others; for example, interdicting a substation may be more costly than interdicting a single transmission line. Under such an alteration, k would be replaced with a total interdiction budget, and the optimization model would find the best interdiction plan for the specified budget.

Model (4.5) is amenable to interpretation; however, it is not immediately tractable as written. First, it is not possible to solve the problem with standard optimization software because some of the variables in the model are attempting to maximize the objective function, while others are attempting to minimize it. Second, the constraints of the model are nonlinear. Developing effective algorithms to solve the interdiction model is central to both the practicality of interdiction modeling and the majority of research in the area.

There are a number of ways of reformulating and solving a model like (4.5) in a tractable fashion. The major methods to gain tractability are:

1. Use a heuristic search to find the critical components (Salmerón et al., 2004).
2. Linearize the products of binary variables and take the dual of the inner problem to obtain a resulting MIP with a single maximization operator (Salmerón et al., 2004).
3. Apply Benders' decomposition (Alvarez, 2004).
4. Rewrite the inner minimization model by forming its optimality (KKT) conditions as constraints (Motto et al., 2005). This method also allows the inner minimization problem to have a different objective function than the outer maximization problem (Bard, 1998), which is sometimes desirable.
5. Develop custom, problem-specific algorithms, which subsequently may generalize to handle other problems (Salmerón et al., 2009).

It is often the custom, problem-specific algorithms that lead to truly large-scale tractability of the interdiction problem—as is the case for interdicting an electric power grid.

Salmerón et al. (2009) develop what they call a global Benders' decomposition algorithm to solve for the most critical components of the electric power grid. The need for such an algorithm arises because the optimal value of the inner minimization is not a concave function in the interdiction variables (or, rather, on the convex hull of their domain). This issue arises frequently in interdiction models. For example, the time to complete the adversary's project, $f(\cdot)$, in the model of Sect. 4.1, is convex on the convex hull of $\mathcal{X}$, yet the interdictor seeks to maximize that function. Such a setup does not naturally lend itself to Benders' decomposition, which, by design, forms an outer linearization of the objective function of a convex program. In some cases, because of the binary nature of the interdiction variables, it is possible to reformulate the inner problem with the interdiction variables instead in the objective function, leading to maximization of a concave function (Cormican et al., 1998; Morton et al., 2007). However, this is not easily done in the case of the inner model in (4.5), largely because of constraint (4.5e). The ability to apply the global Benders' decomposition algorithm of Salmerón et al. hinges on being able to: (a) evaluate the optimal load shedding of the inner problem given δ and (b) form an affine majorizing function of optimal value of the inner minimization, even though it is not concave in δ. Salmerón et al. (2009) show how to do so using properties of the optimal power-flow model under two empirically verified assumptions. This allows us to solve for the most critical components of large-scale instances and yields explicit optimality gaps if the algorithm is terminated prematurely.

4.2.3 Practical Implications and Insights

Practically, a number of aspects of model (4.5) can be altered to yield higher fidelity results. As indicated above, it may be more difficult to disable an entire substation than a single transmission line. This can be reflected by altering constraint (4.5b) to take into account the relative ease of disabling each component. With such a modification, one can derive realistic efficiency curves, measuring the vulnerability of the grid, in load shedding, as a function of an adversary's capability, captured by their budget for interdiction. If the vulnerability curve is relatively flat, the grid can maintain functionality as we increase the number of failed components, or rather the interdiction budget. If the vulnerability curve increases sharply, it indicates a fragile grid in which the load shedding increases sharply with a few failed components.

It is also possible to restrict the set of components that can be interdicted. As described, every component has an associated interdiction variable. However, if it is of interest to consider the vulnerability of the entire grid to the failure of components in a particular geographic region, we can restrict the interdiction model to select only components in a particular region. Another useful restriction involves restricting interdiction to a particular type of component. In this way, we can identify the most critical power generators, for example.

The ability of interdiction algorithms to scale to grids with thousands of components allows us to analyze realistic scenarios on large-scale problems. We can use the interdiction model to answer questions such as: What are the three most critical substations in California? Is there a small set of five to ten components whose failure can cause a large amount of load shedding for a long period of time? And finally, if several candidate plans for electric power grid expansion, or hardening, are proposed, which ones decrease the vulnerability of the grid most effectively?

4.3 Monitoring Our Drinking Water Supply

Following the attacks of September 11, 2001, the US government distributed responsibility of the nation's critical infrastructure to both newly founded and existing federal agencies. The Environmental Protection Agency (EPA) is charged with leading protection of the nation's water supply (Bush, 2003). At about the same time, the US Government Accountability Office identified "distribution systems as among the most vulnerable physical components of a drinking water utility," placing highest priority on the need to develop new technologies to monitor and "quickly detect contaminants in treated drinking water on its way to consumers (GAO, 2003)." These government priorities have directed a decade-long research effort to develop and deploy early warning systems for rapid detection of contaminants in our drinking water.

A central problem in designing a warning system to detect contamination is selecting the best locations for a limited number of water-quality sensors. Historically, optimization models for selecting sensor locations explicitly contained constraints that model water flow (Lee and Deininger, 1992). However, modeling water flow using simulators such as EPANET can produce higher fidelity, physics-based predictions of contaminant flow (Rossman, 2000). Because of the availability of high-fidelity water flow simulations, optimization models switched to exploiting the results of the simulators in selecting sensor locations, instead of modeling water flow through constraints (Ostfeld and Salomons, 2004).

A long-running collaboration between the EPA and operations researchers at Sandia National Laboratories has led to the practical application and deployment of well-designed contamination warning systems. Key to this success was representing the sensor placement problem as a well-known operations research problem—the p-median facility location problem (Berry et al., 2004, 2006). Subsequent and significant improvements to this initial modeling step have led to the development and distribution of the TEVA-SPOT software toolkit, a set of tools to help municipal water utilities locate sensors in their water networks (Berry et al., 2010, 2009; Murray et al., 2009, 2010; Watson et al., 2009).

For the remainder of this section, we describe the key steps to locating sensors that monitor our drinking water supply, focusing on some of the optimization models available in the TEVA-SPOT software toolkit. We provide an intuitive interpretation of the models, the basic mathematical formulation, and the results from such analyses. See Hart and Murray (2010) for a review of a number of different optimization models for placing sensors in water distribution systems.

4.3.1 Locating Sensors to Monitor Drinking Water Networks

A drinking water network can be represented as a set of nodes connected by pipes. The level of resolution of the network can vary from application to application. For example, a single node could represent a single house in some applications, while it may represent an entire neighborhood in others. The water flow throughout the network can be quite complex and is determined by time-dependent demand patterns and operation of pumps and tanks.

Consider a contaminant injected at a single node in the network. The injected contaminant would then move through the network, following a water system's complex flow patterns. If the contaminant flows past an installed sensor, the contaminant may be detected and actions mitigating the contamination can be taken. Formulating an optimization model to locate sensors requires clarifying what makes one placement of sensors preferable to another, and this is complicated by a number of factors including a system's complex flow patterns, questions regarding where the contaminant may be injected, and the inherent stochasticity of sensor equipment and detection events.

The initial models of the sensor placement problem assume perfect sensors and simply seek to maximize *contamination detection coverage* (Lee and Deininger, 1992). With this objective, a node v in the network is considered covered, if a contaminant injection at v is detected at any point in the future by some sensor. While such objectives provide a good starting point for investigation, they can produce somewhat unrealistic results. For example, consider a simple network with two nodes in a line—an upstream node and a downstream node. Suppose that we have to place a single sensor at one of the two nodes. For a contaminant injection at the upstream node, a detection coverage objective would evaluate placing the sensor in either node as equally good because both locations detect the injection. This objective function misses the fact that, in reality, a detection after contaminated water has reached many households is not as valuable as a detection before contaminated water has reached a large segment of the population.

That is why modern sensor placement formulations consider an objective function that minimizes the impact of a contamination event (Berry et al., 2006). The *impact* of a contamination can be defined in terms of the number of people exposed to the contaminant, the key facilities exposed to the contaminant, the length of time of the exposure, or even a combination of such factors (Hart et al., 2012). The sensor placement analysis takes as input a set of contamination scenarios, with each scenario providing a contaminant injection point, injection rate, and length of time for the injection. The analysis can seek to position sensors throughout the network to minimize the *expected impact of contamination*, taken over the provided contamination scenarios (Berry et al., 2006). It is possible that placing sensors to minimize expected impact does not adequately detect a few contamination scenarios with severe impacts. For this reason, we may seek to minimize the *impact of the worst contamination* scenario or to minimize other risk measures, such as the value at risk or the tail-conditional expectation (Watson et al., 2009).

Computing good sensor locations in large water networks, for many contamination scenarios, using meaningful objective functions, as informed by water flow simulations, leads to a complex combinatorial optimization problem. A key insight by Berry et al. (2006) shows how a variant of this problem can be reduced to the well-known p-median problem. This initial step assumes perfect sensors and focuses on minimizing expected impact. Subsequent work expands on the insight to incorporate imperfect sensors and objective functions that incorporate other risk measures (Berry et al., 2009; Watson et al., 2009).

4.3.2 Formulation and Tractability

A basic p-median model for computing good sensor locations can be constructed as follows. Let $\mathcal{A}$ be a set of contamination scenarios. Each scenario $a \in \mathcal{A}$ completely describes a contamination event, including details such as injection point(s), injection rate, start and stop times, type of contaminant, etc. For each scenario, a high-fidelity water flow simulator such as EPANET (Rossman, 2000) is used to compute a time series of the impact of the scenario. Let $d_a(t)$ denote the impact of scenario a at time t after the start of the simulation.

Let V denote the set of potential sensor locations. For this initial model, we assume perfect sensors that detect the contaminant when concentrations exceed a given threshold. For a contamination scenario a and a sensor location j, let γ_{aj} denote the earliest time at which concentrations of the contaminant at location j exceed the detection threshold. Under the perfect sensor assumption, γ_{aj} is the time at which a sensor installed at j sounds an alarm for scenario a. Let $d_{aj} = d(\gamma_{aj})$ be the impact of scenario a if it is first detected by a sensor at location j. Some sensor locations may never detect the contamination scenario. For such locations, we set d_{aj} to be the total impact of the undetected contamination scenario. In reality, if a scenario is not detected by any sensor, it might be detected by another means, such as reported illnesses.

Let α_a denote the probability of encountering contamination scenario a, and suppose we are limited to installing at most p sensors. We can formulate the problem of finding sensor locations that minimize the expected impact of contamination over all scenarios as:

$$\min_{x,s} \quad \sum_{a \in \mathcal{A}} \alpha_a \sum_{j \in V} d_{aj} x_{aj} \tag{4.6a}$$

$$s.t. \quad \sum_{j \in V} x_{aj} = 1, \quad a \in \mathcal{A} \tag{4.6b}$$

$$x_{aj} \leq s_j, \quad a \in \mathcal{A}, j \in V \tag{4.6c}$$

$$\sum_{j \in V} s_j \leq p \tag{4.6d}$$

$$s_j \in \{0,1\}, \quad j \in V \tag{4.6e}$$

$$x_{aj} \in \{0, 1\}, \quad a \in \mathcal{A}, j \in V. \tag{4.6f}$$

The decision variables s_j denote whether location j is chosen for sensor installation, with 1 meaning a sensor is installed and 0 meaning a sensor is not installed. The variable x_{aj} is an auxiliary variable that has value 1 if contamination scenario a is first detected by a sensor at location j, and is 0 otherwise. The objective function in (4.6a) computes the expected impact of contamination over all scenarios, with the inner sum computing the impact of scenario a. Constraint (4.6b) ensures that each scenario is first detected by exactly one sensor; constraint (4.6c) ensures that scenario a can only be detected by a sensor at location j if a sensor is installed at location j; constraint (4.6d) ensures that no more than p sensors are installed; and constraints (4.6e) and (4.6f) ensure binary decision variables.

To gain an intuitive understanding of model (4.6), imagine having a set of five potential sensor locations, $V = \{1, \ldots, 5\}$, and facing a single contamination scenario, a. Suppose we are given an installation plan, for example $s_1 = s_2 = s_3 = 1$ installing sensors in the first three locations, and $s_4 = s_5 = 0$. The x_{aj} variables are simply accounting variables that help us compute the impact of scenario a under the given sensor installation plan. Let j^* be the location with an installed sensor—either 1, 2, or 3 in our example—with minimum impact d_{aj}. When the model computes values for the variables x_{aj}, because of (4.6a)–(4.6c), all x_{aj} are set to zero except x_{aj}^*. Thus, model (4.6) calculates the impact of the scenario under the given sensor installation plan as being equal to $\min_{j|sj = 1} d_{aj}$. This makes intuitive sense since the sensor that sounds the contamination alarm first is the installed sensor giving minimum impact to the contamination scenario. The reasoning in this example also shows that we could relax binary constraint (4.29) in favor of continuous bounds between 0 and 1.

Model (4.6) has the form of the classic p-median facility location problem (see, e.g., Daskin 1995). Sensor locations correspond to facility locations, scenarios correspond to demand points, and impacts d_{aj} correspond to distances between demands and facilities. Recognizing this gives us access to a rich set of tractability improvements based on a large literature devoted to the p-median problem. It is not our purpose here to review such results for the p-median problem. But, we do note that reformulating model (4.6) to aggregate similarly performing sensor locations, dual-based methods employing Lagrangian relaxation, integer-programming based model reductions, and special-purpose heuristics have been widely studied.

Model (4.6) overlooks two important factors in sensor placement. First, sensors are not perfect. They can fail to sense a contaminant (a false negative), and they can alarm when there is no contaminant (a false positive). Berry et al. (2009) show how to incorporate such imperfect sensors into model (4.6). The basic idea of the reformulation is to change the meaning of the variables x_{aj} to the probability that scenario a is first detected by a sensor at location j. These probabilities can be computed by ordering sensor locations in a temporal manner, with locations that have an opportunity to detect the contamination first coming first in the ordering. The probability calculations lead to a nonlinear program, which can be linearized at the cost of a significant increase in the number of decision variables. Nevertheless, the linearized program is able to compute good sensor locations when the sensors are imperfect.

The second issue with model (4.6) is that an objective function minimizing the expected impact over all scenarios can leave some contamination scenarios with extraordinarily high impacts undetected. This is especially a problem if there is reason to believe an adversary could observe the design of our system and exploit it. Even if there is no adversary observing our designs, simply determining scenario probabilities can be extremely difficult. In either case, we may still have interest in the vulnerability of our system to a collection of posited attack scenarios. In this setting, we should minimize the impact of the worst contamination scenario. In other words, we would like to replace objective (4.6a) to change the formulation to

$$\begin{aligned} \min_{x,s} \max_{a \in \mathcal{A}} \quad & \sum_{j \in V} d_{aj} x_{aj} \\ \text{s.t.} \quad & \text{constraints } (6b) - (6f). \end{aligned} \tag{4.7}$$

Model (4.7) has a natural interpretation: First we select sensor locations s; second, an adversary, knowing our sensor locations, selects the worst contamination scenario. Such a model more naturally applies to a terrorist action than does model (4.6). Solving model (4.7) is possible through a standard linearization

$$\begin{aligned} \min_{x,s,y} \quad & y \\ s.t. \quad & \sum_{j \in V} d_{aj} x_{aj} \le y, \quad a \in \mathcal{A} \\ & \text{constraints (4.6b)–(4.6f)}, \end{aligned} \tag{4.8}$$

that is equivalent to the p-center problem. Watson et al. (2009) indicate that it is possible to have both a low expected impact over all scenarios and a low impact for the worst scenario. While we do not detail it here, additional risk measures including value at risk and tail-conditional expectation are explored in Watson et al. (2009).

4.3.3 Practical Implications and Insights

Practical instances of the sensor location problem can be so large that they do not fit in the memory of a typical 32-bit workstation. One example in the literature involves a network with about 12,000 nodes, with sensor locations hedging against 39,000 contamination scenarios. A naive formulation of models (4.6) or (4.7) would require about a half billion variables (Berry et al., 2006; Murray et al., 2009). Even heuristics for sensor placement require on the order of 8 GB of memory to solve such an instance. Through careful reformulations and special purpose algorithms, researchers have been able to find near-optimal solutions to such instances in seconds on a standard laptop computer.

The EPA and Sandia National Laboratories have packaged these methods for computing water sensor locations for municipal water systems in a software package called TEVA-SPOT (Berry et al., 2010), which is available for free download on the Internet (Sandia National Laboratories, 2012). The package has been used to analyze the networks of at least 18 water utilities, and the results have been used to operationally deploy sensors in at least eight utilities. The mean savings from sensor deployment, in terms of reduction in the expected economic impact of a contamination incident if one were to occur, ranged from \$1 to \$33.4 billion with a median of \$5.8 billion. The expected economic impact of the worst contaminations, those in the 95th percentile, dropped by a median of \$19 billion. In more than half of the utilities studied, the expected number of fatalities expected from a contamination dropped by at least 50% (Murray et al., 2009).

The long-term research efforts in water sensor placement have significantly increased the use of operations research in the water resource planning community. Interactions between government, academia, and industry have prompted a realistic mathematical model design and resulted in theoretical, computational, and operational advances. The resulting software continues to be improved and employed by the EPA, with the goal of securing the more than 50,000 water utilities across the USA.

4.4 Securing a Border Against a Nuclear Smuggler

The international atomic energy agency (IAEA) maintains an Illicit Trafficking Database (IAEA, 2012) to which over 100 nation states contribute by reporting events involving illicit trafficking, or other unauthorized possession, of nuclear material and other radioactive material. From 1993 to 2011 over 2,000 such incidents were reported, and about 400 incidents involved criminal activity. During this same time period, 16 cases involved weapons grade material, i.e., highly-enriched uranium (HEU) or plutonium. Some of these seizures involved kilograms of material, and some represented small samples from a larger unsecured stockpile. The IAEA reports that when such information is available, the majority of the cases concerned traffickers seeking financial gain by attempting to sell illicit material. That said, the motives of transporters of illicit nuclear material may change as the material changes hands following its theft and moves along the "supply chain" required to form a weapon. Many of these cases involve perpetrators characterized as being amateurs, but the IAEA reports that some incidents involve organized, professional groups with a history of illicit trafficking in nuclear material. The cases involving HEU and plutonium appear to have originated in Russia or neighboring states, where material was not adequately secured after the fall of the Soviet Union.

The US Department of Homeland Security's Domestic Nuclear Detection Office (DNDO) is charged with developing the global nuclear detection architecture (GNDA). This involves coordination with multiple federal agencies, including the Department of Energy (DOE), the Department of Defense, the Department of State, the Nuclear Regulatory Commission, and coordination with foreign partners. For example, the DOE's National Nuclear Security Administration (NNSA) works with foreign governments to

> deter, detect, and interdict illicit trafficking in nuclear and other radioactive materials across international borders and through the global maritime shipping system. The goal is to reduce the probability of these materials being fashioned into a weapon of mass destruction or a radiological dispersal device ("dirty bomb") to be used against the USA or its key allies and international partners (NNSA, 2012).

There is a strong need for developing better radiation detectors that can sense material like HEU, which can be difficult to detect. At the same time, these detectors should be able to differentiate threats from naturally occurring radioactive material. There is much research in developing effective detectors of radioactive material. That said, there is also important research in how to best deploy and operate these detectors on a large-scale transportation network. Much of the GNDA deployment effort to date, both domestically and abroad, has involved NNSA and DNDO installing radiation portal monitors (RPMs) at seaports, airports, and rail and road border crossings. DNDO also equips Customs and Border Protection officers with mobile detectors and has proposed development of additional mobile detection units, which could be deployed in a surge operation, informed by shorter time-scale intelligence. There are initiatives that seek to secure cities and to deal with difficult challenges such as detecting nuclear smuggling between authorized ports of entry, with small maritime craft, and via general aviation (GAO, 2011).

4.4.1 Locating Radiation Detectors

DNDO has indicated an effort to incorporate increased analytical rigor in its development and analysis of the GNDA (Domestic Nuclear Detection Office, 2012-02). There is a small but growing literature in operations research concerning rigorous analytical models for detecting nuclear material. Wein et al. (2007) propose improvements to an existing spatial deployment of RPMs at a foreign port to increase effectiveness of the system without increasing congestion. Gaukler et al. (2011) and Wein et al. (2006) both employ queueing network models to characterize congestion in a multilayered security system at a seaport, and they seek to optimize the inspection strategy, understanding the tradeoff between detection probability and congestion. For further work on inspection strategies at a single port, see Boros et al. (2009), Madigan et al. (2007), McLay et al. (2011), and Stroud and Saeger (2003). Atkinson et al. (2008) develop a model of a radiation detection system in and around a city, wherein an adversary attempts to get as close as possible to a target in a city center before detonating a nuclear weapon. Cheng et al. (2009) and Hochbaum and Fishbain (2011) analyze mobile distributed detection systems, in which nuclear detectors are mounted on a fleet of many cars—e.g., taxi cabs and/or police cars.

In the remainder of this section, we review a strategic-level model that places RPM detectors at seaports, airports, and rail and road border crossings. The development of this model began as part of the DOE's Second Line of Defense Program (Morton et al., 2007; Pan et al., 2003) and later was coupled with physics-based estimates of detection probabilities and adapted for US ports of entry (Dimitrov et al., 2011). The model we review here is the simplest of a family of models that has been developed. The simple model addresses securing the border of a single country, for example that of Russia or the USA; deals with only stationary detectors; and assumes that both the interdictor and the nuclear smuggler have the same perception of the detection probabilities. Models that relax all of these assumptions have also been developed (Morton et al., 2007; Nehme, 2009; Pan and Morton, 2008; Sullivan et al., 2012).

A key aspect of our model is the transportation network used by the smuggler to move the nuclear material. A smuggler starts at some origin in the network and would like to move to some destination. The transportation network may involve multiple modes of transport; however, our assumption of securing a single country ensures that the smuggler crosses at most one border crossing on his way from the origin to the destination. We restrict attention to installing radiation detectors on the country's legitimate border crossings.

A smuggler may be detected both by indigenous law enforcement, without radiation detectors, and by detectors at border crossings. An intelligent smuggler chooses an origin–destination path to maximize the probability he evades detection, and we assume that he does so knowing the location of radiation detectors. The interdictor does not know the type of smuggler he might face, or the smuggler's origin or destination. We model this lack of complete information as a probability distribution over a range of possible threat scenarios, each specifying a possible smuggler adversary. A threat scenario specifies the smuggler's origin–destination pair; the type of material he smuggles, including its mass, isotopic composition, and geometry; the manner in which that material is shielded, for example by lead of a specified thickness; and the fashion in which the material is transported. Each of these has further detail. For example, the manner in which it is transported can include its position in a rail car or a tractor-trailer container, whether it is a single mass or distributed in the container, and the nature of the accompanying material in the container. All these factors—and more, concerning the type of detector, the algorithm by which it alarms, background radiation from pavement, and whether it has recently rained—contribute to the probability an RPM will detect smuggled material. Subject to resource limits, the interdictor selects sites to install detectors to minimize the system-wide evasion probability.

Following the structure of the models in the three previous sections, the timing of the interdictor's and smuggler's decisions, along with the realization of the threat scenario, is as follows: First, the interdictor installs detectors at a subset of border crossings, subject to a budget constraint. Then, a threat scenario unfolds and the smuggler selects an origin–destination path. The manner in which the smuggler chooses a path is important in determining the best placement of detectors. The model we describe is conservative in that it assumes the smuggler has full knowledge of detector locations and detection probabilities. It is possible to develop models with limited information or different strategies governing the smuggler's behavior. Solutions derived from the conservative model we present have a guaranteed level of performance against more limited adversaries; however, solutions from models from limited adversaries typically do not guarantee performance against an intelligent and informed adversary. While the model we describe specifies a smuggler origin and destination, mathematically, this includes as an important special case a smuggler who optimizes over origin, destination, or both.

4.4.2 Formulation and Tractability

We formulate the model just sketched using the following notation:

Set	
$k \in K$	Border checkpoints
Data	
b	Budget for installing detectors
c_k	Cost of installing detector at k
Random Elements	
$\omega \in \Omega$	Threat scenarios
ϕ^{ω}	Probability mass function on threat scenarios
p_k^{ω}	Evasion probability at k, under ω, when no detector is installed
q_k^{ω}	Evasion probability at k, under ω, when a detector is installed
γ_k^{ω}	Evasion probability on origin–destination path through k, excluding checkpoint k
Decision Variables	
x_k	Binary variable indicating whether (1) or not (0) a detector is installed at k
θ^{ω}	Evasion probability under threat scenario ω

The formulation of the one-country smuggler interdiction model is then:

$$\min_{x,\theta} \quad \sum_{\omega \in \Omega} \phi^{\omega} \theta^{\omega} \tag{4.9a}$$

$$\text{s.t.} \quad \sum_{k \in K} c_k x_k \leq b \tag{4.9b}$$

$$\theta^{\omega} \geq \gamma_k^{\omega} p_k^{\omega} (1 - x_k), \quad k \in K, \omega \in \Omega \tag{4.9c}$$

$$\theta^{\omega} \geq \gamma_k^{\omega} q_k^{\omega} x_k, \quad k \in K, \omega \in \Omega \tag{4.9d}$$

$$x_k \in \{0,1\}, \quad k \in K. \tag{4.9e}$$

Constraints (4.9b) and (4.9e) ensure yes–no detector installation decisions, which satisfy the budget constraint. Constraints (4.9c) and (4.9d), coupled with minimization of the objective function, define the evasion probability for a smuggler, conditional on threat scenario ω, as

$$\theta^{\omega} = \max_{k \in K} \left\{ \gamma_k^{\omega} p_k^{\omega} (1 - x_k)\, ,\, \gamma_k^{\omega} q_k^{\omega} x_k \right\},$$

which encodes the assumption that the smuggler chooses a border crossing to maximize his evasion probability. If the smuggler chooses checkpoint k, then his evasion probability is the product of: (a) his evasion probability from his origin to the checkpoint, (b) his evasion probability from just past the checkpoint to his destination, and (c) his evasion probability through the checkpoint itself. The evasion probability (c) depends on whether a detector is installed at k, and hence is either p_k^{ω} if $x_k = 0$ or q_k^{ω} if $x_k = 1$. The product of the probabilities in (a) and (b) is γ_k^{ω}. The value of γ_k^{ω} can be precomputed by finding maximum evasion probabilities from the origin to each checkpoint, k, and from each checkpoint to the destination, using maximum-reliability-path calculations.

While there are an enormous number of factors that affect the detection probability of an RPM, under mild assumptions, we can aggregate many of these and achieve an equivalent model (Dimitrov et al., 2011). This significantly reduces model complexity. We can further simplify model (??) by replacing constraints (4.33) and (4.34) with

$$\theta^{\omega} \geq r_k^{\omega} (1 - x_k), \quad k \in K, \omega \in \Omega, \tag{4.10}$$

where $r_k^{\omega} = \max(\gamma_k^{\omega} p_k^{\omega} - q_{\max}^{\omega}, 0)$ and $q_{\max}^{\omega} = \max_{k \in K} \gamma_k^{\omega} q_k^{\omega}$. The resulting model is equivalent, with its optimal value differing from that of model (4.9) by the constant $\sum_{\omega \in \Omega} \phi^{\omega} q_{\max}^{\omega}$.

Still, the linear-programming relaxation of model (4.9) is so weak that the required computational effort to solve realistically sized instances is prohibitive. We can gain computational traction in model (4.9) by observing that constraints (4.10) have the form of the so-called *mixing inequalities* (see Miller and Wolsey 2003 and references therein). This opens two computationally promising avenues. One is rooted in using an exponentially sized class of valid inequalities (Günlük and Pochet, 2001; Pochet and Wolsey, 1994), which can be separated in polynomial time by solving an appropriately defined shortest-path problem. This avenue has been pursued for model (4.9) (Morton et al., 2007) and for variants of model (4.9) where the smuggler and interdictor have differing perceptions of evasion probabilities, respectively (Pan and Morton, 2008; Sullivan et al., 2012). The second avenue is to use a so-called *extended formulation* for the mixing inequalities (Miller and Wolsey, 2003; Pochet and Wolsey, 1994) to tighten the formulation. This has been developed for model (4.9), and we describe this extended formulation next (Nehme and Morton, 2009).

Thinking from smuggler ω's perspective, we sort the transformed evasion probabilities: $r^{\omega}_{k(1,\,\omega)} \geq r^{\omega}_{k(2,\,\omega)} \geq \cdots \geq r^{\omega}_{k(\,|\,K\,|\,,\,\omega)}$. Here, $k(i,\,\omega)$ denotes smuggler ω's i-th best checkpoint. We define $\Delta_{k(i,\,\omega)}{}^{\omega} = r^{\omega}_{k(i,\,\omega)} - r^{\omega}_{k(i\,+\,1,\,\omega)}$, $i = 1, \ldots, |\,K\,|$, as the reward the interdictor collects by forcing smuggler ω from his i-th to his $(i + 1)$-st best checkpoint. And, we introduce decision variable $u_k{}^{\omega}$, which takes value 1 if smuggler ω is forced to a checkpoint lower than k on this sorted list. With boundary conditions $r^{\omega}_{k(\,|\,K\,|\,+\,1,\,\omega)} = 0$ and $u^{\omega}_{k(0,\,\omega)} = 1$ we have:

$$\theta^{\omega} = \sum_{i=1}^{|K|} r^{\omega}_{k(i,\omega)}\left(u^{\omega}_{k(i-1,\omega)} - u^{\omega}_{k(i,\omega)}\right) = r^{\omega}_{k(1,\omega)} + \sum_{i=1}^{|K|} \underbrace{\left(r^{\omega}_{k(i+1,\omega)} - r^{\omega}_{k(i,\omega)}\right)}_{-\Delta^{\omega}_{k(i,\omega)}} u^{\omega}_{k(i,\omega)}. \tag{4.11}$$

Upon substituting (4.11) we have the following reformulation of model (4.9):

$$\max_{x,u} \quad \sum_{\omega\in\Omega}\sum_{k\in K} \phi^{\omega}\Delta^{\omega}_k u^{\omega}_k \tag{4.12a}$$

$$\text{s.t.} \quad \sum_{k\in K} c_k x_k \leq b \tag{4.12b}$$

$$u^{\omega}_k \leq x_k, \quad k \in K, \omega \in \Omega \tag{4.12c}$$

$$u^{\omega}_{k(i,\omega)} \leq u^{\omega}_{k(i-1,\omega)}, \quad i = 2, \ldots, |K|, \omega \in \Omega \tag{4.12d}$$

$$0 \leq u^{\omega}_k \leq 1, \quad k \in K, \omega \in \Omega \tag{4.12e}$$

$$x_k \in \{0, 1\}, \quad k \in K. \tag{4.12f}$$

The constraints of model (4.12) capture those of model (4.9) with the addition of constraints (4.12c)–(4.12e) to define u^{ω}_k; i.e., they allow reward Δ^{ω}_k to be collected only if a detector is installed at checkpoint k and at all the checkpoints that smuggler ω ranks above k. The variable u^{ω}_k is naturally binary, given that we require x_k to be binary. Model (4.12) has a much tighter linear-programming relaxation than that of model (4.9), allowing us to solve large instances.

4.4.3 Practical Implications and Insights

Rather than viewing constraint (4.12b) as a hard budget constraint, it typically makes sense to study the trade-off between system performance—in this case, the probability we detect a smuggler—and the cost of the associated system design. To do so, we can solve model (4.12) parametrically in the budget b to obtain the set of Pareto efficient solutions. If we modify model (4.12) by removing constraint (4.12b) and instead maximizing the objective function

$$\sum_{\omega \in \Omega} \sum_{k \in K} \phi^{\omega} \Delta_k^{\omega} u_k^{\omega} - \lambda \sum_{k \in K} c_k x_k,$$

where λ parametrically ranges over positive values, we can obtain a subset of Pareto efficient solutions. Specifically, we obtain those that are extreme points of the concave envelope of the efficient frontier (Kuhn and Tucker, 1951; Nehme and Morton, 2010). Note that when relaxing the model in this way, so that we have a soft budget constraint, model (4.12)'s constraint set has a dual-network structure. The constraint matrix is totally unimodular, as each structural constraint has one + 1 and − 1.

The relaxed model, with the soft budget constraint, has the form of the *selection problem* of Balinski (1970) and Rhys (1970). This leads to a very special structure of the extreme point solutions of the concave envelope of the efficient frontier. In particular, these solutions are *nested* (Hochbaum, 2009; Nehme and Morton, 2010; Witzgall and Saunders, 1988); i.e., if x^* (b) denotes checkpoints which receive detectors under budget b for one such extreme point and x^* (b') for an extreme point under a larger budget, then x^* $(b) \leq x^*$ (b') in the vector sense. In words, this notion of nestedness means that the optimal set of checkpoints to receive detectors at budget b is a subset of those at a larger budget b'. This has important practical implications because usually the border, or another system we seek to protect, is incrementally hardened over time as additional funds become available. It is typically impossible, or too expensive, to completely redesign the system as the budget grows. This result yields budget increments at which optimal solutions are naturally nested.

Additional, geographic structure of optimal solutions to model (4.12) exists as we parametrically range the budget, b (Dimitrov et al., 2011). In particular, as the budget grows, the checkpoints that receive detectors fall in geographic clusters. In model instances for installing detectors on the land border crossings of the contiguous US, four geographic clusters emerge: crossings east of Big Bend in Texas, the remaining crossings on the US–Mexico border, crossings in the Great Lakes region and the rest of the northeast, and crossings west of the Great Lakes. The reason for this structure in optimal solutions is as follows: If we are dealing with an intelligent and well-informed smuggler, then installing detectors at only a subset of nearly identical border crossings does not improve our ability to detect the smuggler. Instead, we must equip all checkpoints in a geographic cluster in order to force the smugger to select an alternate path with lower evasion probability.

4.5 Discussion and Conclusions

The four applications discussed above—delaying a nuclear weapons project, assessing vulnerabilities in the electric power grid, detecting contaminants in drinking water, and securing our border against nuclear smugglers—exemplify the utility and the development of interdiction modeling. Analyses using interdiction models have made important contributions at multiple levels of government. They can be used to analyze and harden our critical infrastructure systems as well as to look for vulnerabilities in an adversary's system. Standard pathways for developing interdiction models are a useful first step in analysis, often delivering key insights. In addition, thoughtful, problem-specific interdiction models and optimization methods can elevate the interdiction approach to directly applicable, large-scale settings.

Sometimes, as is the situation for some of the case studies we present, it may take a decade of research and a sequence of insights into a problem to develop the special-purpose methods required to deliver specific and timely guidance on large-scale interdiction problems. Building on initial, stylized models, such an effort can have a marked operational impact. Success can also depend on persistence in delivering the insights from analyzing interdiction models.

Having an impact in practice can further hinge on putting forward compelling arguments, perhaps even based on detailed analysis, showing that less principled approaches to interdiction can yield inferior results, to potentially devastating effect. The two foremost approaches that we categorize as being less principled involve: (a) ignoring the distinction between an intentional attack and a random disruption and (b) ignoring the underlying system. As we discuss above, there is a rich literature on assessing the reliability of a system to random component failure. However, in making a modeling error of type (a), we presume our adversaries will behave similarly. There is ample evidence that this is simply not the case, particularly when our adversaries have the will and means to become well informed as to our system's design, and defenses, and when they seek to inflict maximum damage.

Modeling errors of type (b) are all too pervasive in practice, and even in our literature. In a typical such setting, an analyst develops a measure of an individual component's value. The analyst then "scores" each of the components in the system and sorts to obtain a priority list for components that should be interdicted, or hardened against interdiction. In interdicting a maximum-flow network or in interdicting a shortest-path network, this amounts to forming a sorted list of arcs based on their capacities or lengths. This ignores the fact that a system's performance can depend in subtle, and sometimes surprising, ways on the manner in which the components interact and on key subsets of components, as opposed to individual components. That such subtleties and surprises emerge from our models with regularity is well recognized in operations research. We should not forget this *raison d'être* when seeking to understand the vulnerability of our systems to intentional attack.

Because of their utility, interdiction models have already become a standard element of educational curricula in many operations research programs. Sometimes, interdiction modeling is a part of advanced courses on optimization, and sometimes it is simply included in basic, required courses on network modeling. Giving future operations researchers a good understanding of the principles of interdiction modeling, contrasting interdiction with less suitable approaches, and teaching the basic modeling techniques and computational tools for interdiction, ensures our ability to effectively detect vulnerabilities in the systems we build and uncover such vulnerabilities in our adversaries' systems.

Acknowledgements The authors thank Regan Murray, Javier Samerón, Jean-Paul Watson, and Kevin Wood whose thoughtful comments improved this chapter. This work has been supported by the National Science Foundation through grants CMMI-0653916 and CMMI-0800676, the Defense Threat Reduction Agency through grant HDTRA1-08-1-0029, and the US Department of Homeland Security under Grant Award Number 2008-DN-077-ARI021-05. The views and conclusions contained in this document are those of the authors and should not be interpreted as necessarily representing the official policies, either expressed or implied, of the US Department of Homeland Security.

References

Albert R, Albert I, Nakarado GL (2004) Structural vulnerability of the North American power grid. Phys Rev E 69(2):025103+

Alvarez R (2004) Interdicting electrical power grids. Master's thesis, Naval Postgraduate School, Monterey, CA

Atkinson MP, Cao Z, Wein LM (2008) Optimal stopping analysis of a radiation detection system to protect cities from a nuclear terrorist attack. Risk Anal 28:353–371

Balinski ML (1970) On a selection problem. Manag Sci 17:230–231

Bard JF (1998) Practical bilevel optimization: algorithms and applications. Kluwer, Boston

Barron J (2003) Power surge blacks out northeast. The New York Times, August 15

Berry J, Hart WE, Phillips CA, Uber J (2004) A general integer-programming-based framework for sensor placement in municipal water networks. In: Sehlke G, Hayes DF, Stevens DK (eds) Proceedings of the World Water and Environmental Resources Congress 2004, Salt Lake City, Utah, United States. June 27–July 1

Berry J, Hart WE, Phillips CA, Uber JG, Watson J (2006) Sensor placement in municipal water networks with temporal integer programming models. J Water Resour Plann Manag 132(4):218–224

Berry J, Carr RD, Hart WE, Leung VJ, Phillips CA, Watson J (2009) Designing contamination warning systems for municipal water networks using imperfect sensors. J Water Resour Plann Manag 135(4):253–263

Berry JW, Boman E, Riesen LA, Hart WE, Phillips CA, Watson J-P, Murray R (2010) User's manual: TEVA-SPOT toolkit version 2.4. Technical Report EPA/600/R-08/041, National Homeland Security Research Center, Office of Research and Development, U.S. Environmental Protection Agency

Borger J (2012) Who is responsible for the Iran nuclear scientists attacks? The Guardian, January 12

Boros E, Fedzhora L, Kantor PB, Saeger KJ, Stroud P (2009) Large scale LP model for finding optimal container inspection strategies. Naval Res Logist 56:404–420

Broad WJ, Markoff J, Sanger DE (2011) Israeli test on worm called crucial in Iran nuclear delay. The New York Times, January 15

Brown GG, Carlyle WM, Harney R, Skroch E, Wood RK (2009) Interdicting a nuclear-weapons project. Oper Res 57:866–877

Bush GW (2003) Subject: critical infrastructure identification, prioritization, and protection. Homeland Security Presidential Directive HSPD-7. http://www.dhs.gov/xabout/laws/gc_1214597989952.shtm. Accessed 9 Jul 2011

Chassin D, Posse C (2005) Evaluating North American electric grid reliability using the Barabási–Albert network model. Phys A Stat Mech Appl 355(2–4):667–677

Cheng J, Xie M, Roberts F (2009) Design and deployment of a mobile sensor network for the surveillance of nuclear materials in metropolitan areas. In: Proceedings of 15th international conference on reliability and quality of design (ISSAT09), San Francisco, California, USA. August 6–8

Cormican K, Morton DP, Wood RK (1998) Stochastic network interdiction. Oper Res 46:184–197

Daskin MS (1995) Network and discrete location: models, algorithms, and applications. Wiley, New York

Dimitrov NB, Michalopoulos D, Morton DP, Nehme MV, Pan F, Popova E, Schneider EA, Thoreson GG (2011) Network deployment of radiation detectors with physics-based detection probability calculations. Ann Oper Res 187:207–228

Domestic Nuclear Detection Office (2012) The last line of defense: federal, state, and local efforts to prevent nuclear and radiological terrorism within the United States. DNDO Director Warren Stern, before the House Committee on Homeland Security, Subcommittee on Cybersecurity, Infrastructure Protection, and Security Technologies, July 25, 2011. http://www.dhs.gov/ynews/testimony/20110726-stern-last-line-of-defense.shtm. Accessed 11 Feb 2012

Espiritu J, Coit D, Prakash U (2007) Component criticality importance measures for the power industry. Electr Power Syst Res 77(5–6):407–420

GE Energy for The National Renewable Energy Laboratory (2010) Western wind and solar integration study. The National Renewable Energy Laboratory Report NREL/SR-550-47434

United States General Accounting Office (GAO) (2003) Drinking water: Experts' views on how future federal funding can best be spent to improve security. Technical Report GAO-04-29. Report to the Committee on Environment and Public Works, U.S. Senate

United States General Accounting Office (GAO) (2011) Combating nuclear smuggling: DHS has developed a strategic plan for its global nuclear detection architecture, but gaps remain. Technical Report GAO-11-869T. Testimony Before the Subcommittee on Cybersecurity, Infrastructure Protection, and Security Technologies, U.S. House of Representatives

Gaukler GM, Li C, Cannaday R, Chirayath SS, Ding Y (2011) Detecting nuclear materials smuggling: using radiography to improve container inspection policies. Ann Oper Res 187:65–87

Günlük O, Pochet Y (2001) Mixing MIR inequalities for mixed integer programs. Math Program 90:429–458

Harney R, Brown GG, Carlyle WM, Skroch E, Wood RK (2006) Anatomy of a project to produce a first nuclear weapon. Sci Global Secur 14:163–182

Hart WE, Murray R (2010) A review of sensor placement strategies for contamination warning systems. J Water Resour Plann Manag 136(6):611–619

Hart WE, Berry JW, Murray R, Phillips CA, Riesen LA, Watson J (2012) SPOT: A sensor placement optimization toolkit for drinking water contaminant warning system design. In: Kabbes KC, (ed) Proceedings of the 2007 World Environmental and Water Resources Congress, Tampa, Florida, May 15–19, 2007. http://cfpub.epa.gov/si/si_public_record_report.cfm?dirEntryId=166528. Accessed 12 Mar 2012

Hochbaum DS (2009) Dynamic evolution of economically preferred facilities. Eur J Oper Res 193:649–659

Hochbaum DS, Fishbain B (2011) Nuclear threat detection with mobile distributed sensor networks. Ann Oper Res 187:45–638

International Atomic Energy Agency (2012) Fact sheet: IAEA Illicit Trafficking Database (ITDB). http://www-ns.iaea.org/security/itdb.asp. Accessed 11 Feb 2012

Keizer G (2010) Is Stuxnet the 'best' malware ever? Computerworld, September 16

Kessler G, Wright R (2007) Israel, U.S. shared data on suspected nuclear site. Washington Post, September 21

Kuhn HW, Tucker AW (1951) Nonlinear programming. In: Proceedings of the 2nd Berkeley symposium on mathematical statistics and probability, University of California Press. Berkeley, California, pp 481–492

Lee BH, Deininger RA (1992) Optimal locations of monitoring stations in water distribution system. J Environ Eng 118(1):4–16

Madigan D, Mittal S, Roberts F (2007) Sequential decision making algorithms for port of entry inspection: overcoming computational challenges. In: Muresan G, Altiok T, Melamed B, Zeng D (eds) Proceedings of the 2007 IEEE Intelligence and Security Informatics Conference, New Brunswick, New Jersey, USA. May 23–24, pp 1–7

McGoldrick F (2011) Limiting transfers of enrichment and reprocessing technology: issues, constraints, options. Report for Belfer Center for Science and International Affairs, Harvard Kennedy School, Cambridge, MA

McLay LA, Lloyd JD, Niman E (2011) Interdicting nuclear material on cargo containers using knapsack problem models. Ann Oper Res 187:185–205

Mili L, Qiu Q, Phadke AG (2004) Risk assessment of catastrophic failures in electric power systems. Int J Crit Infrastruct 1(1):38–63

Miller AJ, Wolsey LA (2003) Tight formulations for some simple mixed integer programs and convex objective integer programs. Math Program 98:73–88

Morton DP, Pan F, Saeger KJ (2007) Models for nuclear smuggling interdiction. IIE Trans Oper Eng 38:3–14

Motto AL, Arroyo JM, Galiana FD (2005) A mixed-integer LP procedure for the analysis of electric grid security under disruptive threat. IEEE Trans Power Syst 20(3):1357–1365

Murray R, Hart WE, Phillips CA, Berry J, Boman EG, Carr RD, Riesen LA, Watson J, Haxton T, Herrmann JG, Janke R, Gray G, Taxon T, Uber JG, Morley KM (2009) US Environmental Protection Agency uses operations research to reduce contamination risks in drinking water. Interfaces 39(1):57–68

Murray R, Haxton T, Janke R, Hart WE, Berry J, Phillips C (2010) Sensor network design for drinking water contamination warning systems: A compendium of research results and case studies using the TEVA-SPOT software. Technical Report EPA/600/R-09/141, National Homeland Security Research Center, Office of Research and Development, U.S. Environmental Protection Agency

National Energy Policy Development Group (2001) National Energy Policy: Report of the National Energy Policy Development Group. U.S. Government Printing Office

National Nuclear Security Administration (2012) Fact sheet: NNSA's Second Line of Defense Program. Released 1 Feb 2011. https://nnsa.energy.gov/mediaroom/factsheets/nnsassecondlineofdefenseprogram. Accessed 1 Feb 2012

Nehme MV (2009) Two-person games for stochastic network interdiction: models, methods, and complexities. Ph.D. thesis, The University of Texas at Austin

Nehme MV, Morton DP (2009) Tightening a network interdiction model. In: Proceedings of the 2009 Industrial Engineering Research Conference, Miami, Florida, May 30 – June 3

Nehme MV, Morton DP (2010) Efficient nested solutions of the bipartite network interdiction problem. In: Johnson A, Miller J (eds) Proceedings of the 2010 Industrial Engineering Research Conference, Cancun, Mexico. June, 2010

O'Brien JJ (1969) Scheduling handbook. McGraw-Hill Book Company, New York

Office of Technology Assessment U.S. Congress (1990) Physical vulnerability of electric systems to natural disasters and sabotage. U.S. Government Printing Office. OTA-E-453

Ostfeld A, Salomons E (2004) Optimal layout of early warning detection stations for water distribution systems security. J Water Resour Plann Manag 130(5):377–385

Pan F, Morton DP (2008) Minimizing a stochastic maximum-reliability path. Networks 52:111–119

Pan F, Charlton W, Morton DP (2003) Interdicting smuggled nuclear material. In: Woodruff DL (ed) Network interdiction and stochastic integer programming. Kluwer, Boston, pp 1–20

Pochet Y, Wolsey LA (1994) Polyhedra for lotsizing with Wagner-Whitin costs. Math Program 67:297–323

Qiang Q, Nagurney A (2008) A unified network performance measure with importance identification and the ranking of network components. Optim Lett 2(1):127–142

Rausand M, Høyland A (2003) System reliability theory: models, statistical methods, and applications, 2nd edn. Wiley, Hoboken

Reed BK (1994) Models for proliferation interdiction response analysis. Master's thesis, Naval Postgraduate School, Monterey, CA

Rhys JMW (1970) A selection problem of shared fixed costs and network flows. Manag Sci 17:200–207

Rossman LA (2000) EPANET 2: Users manual. Technical Report EPA/600/R-00/057, United States Environmental Protection Agency

Salmerón J, Wood K, Baldick R (2004) Analysis of electric grid security under terrorist threat. IEEE Trans Power Syst 19(2):905–912

Salmerón J, Wood K, and Baldick R (2004) Optimizing electric grid design under asymmetric threat (II). Technical Report NPS-OR-04-001, Naval Postgraduate School. Prepared for U.S. Department of Justice, Office of Justice Programs and Office of Domestic Preparedness

Salmerón J, Wood K, Baldick R (2009) Worst-case interdiction analysis of large-scale electric power grids. IEEE Trans Power Syst 24(1):96–104

Sandia National Laboratories (2012) TEVA-SPOT Toolkit: A sensor placement optimization tool for water security. https://software.sandia.gov/trac/spot. Accessed 17 Mar 2012

Shtub A, Bard JF, Globerson S (2005) Project management: processes, methodologies, and economics. Prentice Hall, Upper Saddle River

Stroud PD, Saeger KJ (2003) Enumeration of increasing Boolean expressions and alternative digraph implementations for diagnostic applications. In: Proceedings volume IV, computer, communication and control technologies, pp 328–333

Sullivan KM, Morton DP, Pan F, Smith JC (2012) Interdicting stochastic evasion paths with asymmetric information on bipartite networks. Naval Res Logist (under revision)

Talukdar S, Apt J, Ilic M, Lave L, Morgan M (2003) Cascading failures: survival versus prevention. Electr J 16(9):25–31

Tavner PJ, Xiang J, Spinato F (2007) Reliability analysis for wind turbines. Wind Energy 10 (1):1–18

Watson J, Murray R, Hart WE (2009) Formulation and optimization of robust sensor placement problems for drinking water contamination warning systems. J Infrastruct Syst 15(4):330+

Wein LM, Wilkins AH, Baveja M, Flynn SE (2006) Preventing the importation of illicit nuclear materials in shipping containers. Risk Anal 26:1377–1393

Wein LM, Liu Y, Cao Z, Flynn SE (2007) The optimal spatiotemporal deployment of radiation portal monitors can improve nuclear detection at overseas ports. Sci Global Secur 15:211–233

Witzgall CJ, Saunders PB (1988) Electronic mail and the "locator's" dilemma. In: Ringeisen RD, Roberts FS (eds) Applications of discrete mathematics. SIAM, Philadelphia, pp 65–84

Chapter 5
Time Discrepant Shipments in Manifest Data

James Abello, Mikey Chen, and Neel Parikh

Abstract Manifest data is a log of container shipments from foreign lading ports to U.S. unlading ports. We provide several time varying network-based representations of this data in order to extract its most "discrepant" port pairs and contents patterns. We treat this time varying network representation as a combinatorial set system and use its discrepancy and firing rate (Abello et al. (2010) Detecting Novel Discrepancies in Communications Networks, International Conference on Data Mining, ICDM 2010: 8–17, Sydney, Australia and Chazelle (2000) The Discrepancy Method: Randomness and Complexity, Cambridge University Press) as the main statistics to track the most "salient" network elements. The output of the entire process is a "fossil" sub-network that encodes those port pairs and contents that exhibit unusual time varying patterns. It is expected that substantial deviations from these patterns will be useful triggers for further content inspections. The applicability of the proposed techniques is not limited to manifest data.

5.1 Introduction

We obtained manifest data of foreign shipments to U.S. ports from the U.S. Customs and Border Protection Agency (ASFOI.txt 2009). A *manifest shipment* is a logical structure containing information regarding containers shipped between an associated unique pair of ports. Each manifest shipment includes information about one or more containers. Containers may have a free textual description of their contents which usually consist of several packages (see Appendix 1). Each

J. Abello (✉) • M. Chen • N. Parikh
DIMACS Center, Rutgers University, DIMACS Core Building, Room 423,
Busch Campus, 96 Frelinghuysen Road, Piscataway, NJ 08854-8018, USA
e-mail: abello@dimacs.rutgers.edu

J.W. Herrmann (ed.), *Handbook of Operations Research for Homeland Security*, International Series in Operations Research & Management Science 183,
DOI 10.1007/978-1-4614-5278-2_5,

manifest shipment consists of a consecutive collection of fixed format manifest data lines (each with 278 characters). The number and layout of data lines per shipment vary. Each line has a beginning indicator character that corresponds to different layout record types. The different layouts include information about general billing, container type and contents, shipper, consignee, party to notify, hazardous materials, and universal standard international symbols representing contents marks (mrks-nbrs). We parse each line according to its layout type and use meta field length definitions to obtain 114 variables, among which 8 are numerical, 20 are categorical, and 76 are verbiage (see Appendix 1 for a sample of the original data). In the sample that we use for our experimentation, we were provided shipment data among 308 ports. There were 19,038 shipment content descriptions of 32,337 containers that used 17,199 keywords.

The goal of this work was to identify the main underlying time varying characteristics of shipment content patterns. The expectation is that substantial deviations from these patterns can be used as triggers for further content inspections. We adapt the data model proposed by Abello et al. (2010) to transform the raw manifest data into a sequence of time stamps, each with a corresponding set of events that occur between related data entities. We explain in Sect. 5.2 how the raw manifest data is transformed and combined into a collection of time varying graphs. The central idea is to associate to each port pair a weighted and time ordered contents multi-set. In this way, each triple <port, port, contents_descriptor> gets assigned a time varying firing rate.[1] In Sect. 5.3, we describe how combinatorial discrepancy (Chazelle 2000) can be applied to "extract" the most "salient" <port, port, contents> triples. Section 5.4 describes how these triples can be visualized and analyzed in order to extract the most time varying atypical <port, port, contents> triples. For a given time window, the union of these time varying "salient" triples constitutes a historical sketch of the manifest data. This historical sketch is, to an extent, an analog of what archeologists call a fossil. We call this specially extracted triple collection a "data fossil" discussed in Sect. 5.5. Section 5.6 reports some of our findings and conclusions and points out some other potential applications of our techniques.

5.2 Similarity Coupling of Shipments via Contents Vectors

5.2.1 Associating a Content Vector to Each Active Port Pair

For a particular collection of manifest data records, stamped with the same U.S. unlading date, we associate to each pair of ports $u' =$ (origination, destination) a contents vector $C_{u'}$. $C_{u'}$ encodes information about the contents shipped from an

[1] See Sect. 5.3.

originating foreign port to a U.S. destination port in all shipments present in the data. Next, we couple two port pairs $u' =$ (origin 1, destination 1) and $v' =$ (origin 2, destination 2) via a modified dot product between their content vectors $C_{u'}$ and $C_{v'}$:

$$w(u', v') = \frac{C_{u'} \bullet C_{v'}}{C_{u'}^2 + C_{v'}^2 - C_{u'} \bullet C_{v'}}.$$

This provides us with a global weighted graph where the vertices are port pairs and the similarity between two pairs is defined as a function of the dot product of their corresponding content vectors. We call this global weighted virtual topology Daily_PortPairs_Coupling. It is specified by two files. One describes the virtual topology and the other associates to each port pair a vector label. This vector label not only disambiguates the port pairs, but it also associates with each port pair a multi-set of contents. The overall shipment contents are organized in a hierarchy of Contents_Descriptors (CD for short). This hierarchy is obtained by an iterative process that enlarges a "current" set of key words (minus stop words) depending on their matching with the textual description of the overall container contents. By clustering this representation according to a variety of similarity measures, we obtain typical patterns of port pairs traffic with respect to their shipment contents. The great majority of patterns consist of a *peripheral independent subset of vertices IP* connected to a disjoint *"central" subset of vertices C of high edge density*. Figures 5.1 and 5.2 illustrate some of these typical manifest data patterns where the central set C is a clique.

A variety of other cliques were detected when clustering port pairs according to the contents they shipped between them. The largest of such clique corresponded to 15 port pairs shipping products in the arts-crafts category originating from ports in Panama, Brazil, Argentina, Spain, Italy, France, and Australia.

Motivated by the previous findings we extended the notion of contents vector to an entire day of manifest data. Surprisingly, all vectors corresponding to the same day of the week during a month of manifest data were grouped together by the Hellinger distance.

5.2.2 Daily Manifest Shipment Content Vectors and Hellinger Distance

Given a date x of Manifest Data, consider a vector C_x with entries labeled by the content descriptors appearing as leaves in the Content_Descriptors hierarchy. Each entry $C_x(d)$ of this vector encodes the number of "items" of contents of type d shipped on x from all foreign ports to all U.S. ports. This collection of vectors for a consecutive interval T of days is a very high level aggregate description of the contents shipments from all foreign ports to all US ports. It can be viewed as a matrix M with $|T|$ rows and |Content_Descriptors| columns. We normalize each column by

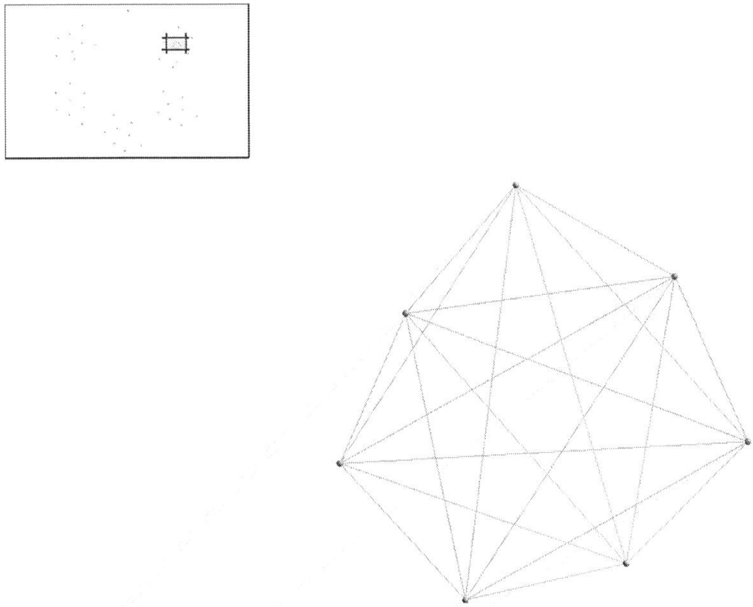

Fig. 5.1 The depicted clique is automatically extracted from the Daily_PortPairs_Coupling graph. Each vertex corresponds to a port pair. The *vertex color* encodes the overall number of pieces shipped between the two ports. The *edges* encode the contents similarity between the corresponding two port pairs. Five nodes represent shipments originating from Rotterdam (Netherlands) and Anvers (Belgium). The remaining two nodes correspond to shipments from Veracruz (Mexico) and Port Bustamante (Jamaica). In the month of data used for our experimentation, home and gardening products were the most common dominant goods being shipped among all these port pairs

its column sum and then normalize the resulting row by its row sum. We compute next the Hellinger distance $h(P,Q)$ between every pair of row vectors P and Q, which is the 2-norm of the square roots (entry-wise) of P and Q (Lin 1991):

$$h(P,Q) = \frac{1}{\sqrt{2}}\left\|\sqrt{P} - \sqrt{Q}\right\|_2.$$

The matrix formed by this collection of pair-wise Hellinger distances is used to compute a minimum spanning tree that we call the Hellinger Tree. It is a remarkable fact that all the vectors corresponding to the same day of different weeks are grouped together in the Hellinger Tree. Figure 5.3 illustrates this finding for one month of manifest data.

Up to this point we have discussed useful global aggregates that allow us to compare port pairs shipments and even entire days of manifest data with respect to their contents descriptors. In the next section, we discuss how we compared port

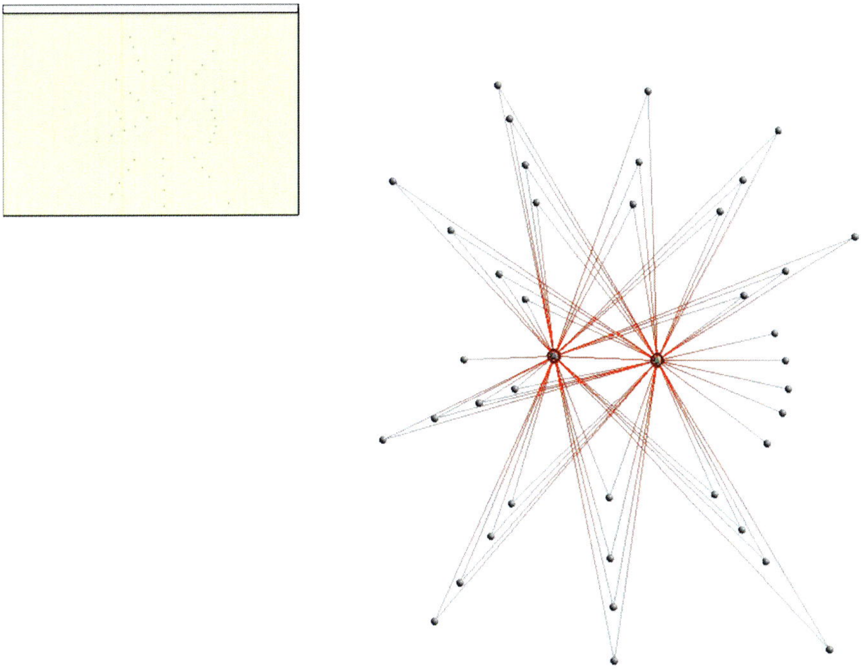

Fig. 5.2 The central set in this pattern consists of just two vertices connected by an edge. They correspond to the two port pairs (Salalah: Oman, Newark: NJ) and (Port Sweeten: Malaysia, Savanna: Georgia). Their shipment patterns dominate all the other port pairs in the peripheral set even though the port pairs in the peripheral set are not similar among each other. An alternative view is that the multi-set of good shipped among the central two vertices is approximately equal to the disjoint union of the multi-sets shipped by the peripheral vertices

pairs at the lowest level of granularity, i.e., for a given content descriptor *cd*, how can we encode the traffic patterns of all the *cd* shipments between a port pair (u, v) and another pair (u', v') during a given time interval T. The fundamental idea is to identify each triple $\langle u, v, cd \rangle$ with a weighted set of time stamps and to treat the entire manifest data as a combinatorial weighted time system. This encoding will allow us to extract the most "discrepant" time varying patterns.

5.3 Shipments Traffic and Time Set Systems

To encode time varying shipment patterns, we view the entire manifest data as a weighted multi-set system: one multi-set $M_{u,v}$ per port pair (u, v). Each multi-set $M_{u,v}$ is represented as a matrix with dimensions time and contents descriptors. Each entry $M_{u,v}(t, cd)$ records the percentage of goods in category *cd* shipped from port u to port v at time t. Alternatively, each column t encodes the weighted set of "items" shipped at time t from foreign port u to U.S. port v. With these conventions, a

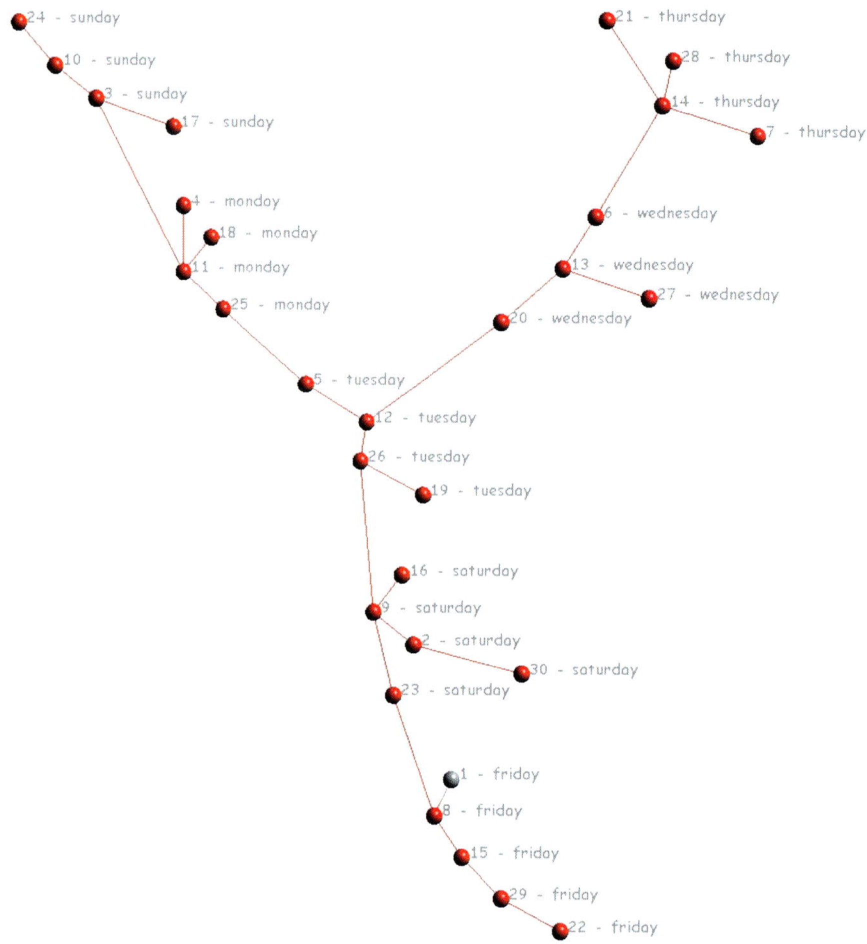

Fig. 5.3 The Hellinger Minimum Spanning Tree of one month of Manifest Data. Every vertex represents one day from January 30, 2009, up to February 28, 2009. The number next to the vertex corresponds to the actual date. The weight of an edge (u, v) is equal to the Hellinger distance between the two corresponding content vectors Cu and Cv

row labeled cd is nothing but a time series keeping track of the shipment activity from port u to port v of contents labeled cd. Dually, for any fixed type of contents cd the set of all quadruples $\langle u, v, t, cd \rangle$ such that $M_{u,v}(t, cd)$ is nonzero defines a time varying bipartite graph encoding all of the shipments of contents of type cd from foreign ports to U.S. ports for t varying in a particular time interval. This bipartite graph can be viewed as a matrix of port pairs and time where each entry encodes the cd shipments "volume" from one port to another at a particular time t. Notice that we use percentages rather than absolute volume counts because different content types have quite different volume units (e.g., cars vs. paper rolls). A natural question is: which port pairs may be considered "salient" with respect to their time varying shipments of contents labeled *cd*? Certainly those pairs whose shipment activity is above the average should be considered "salient." A more refined answer is that "salient" port pairs are those whose *shipment activity patterns* deviate substantially from the *overall activity pattern* of shipments with respect to a particular type of contents. As an illustration consider the following not so hypothetical statement: "During the last decade, it has been observed that most shipments of electronics to the U.S. originated in certain subset of foreign ports during a particular month. However, quite recently a substantial shipment volume is originating from a different port set." At a very high level of granularity, this type of question can be answered by using a well-designed set of queries formulated in standard query languages applied to appropriate historical data. However, in a variety of streaming settings, we want to be able to provide summaries of "salient" pairs of entities that exchange commodities at different time varying rates. For this setting, combinatorial discrepancy (Abello et al. 2010) has been proposed as a complementary tool to more traditional analysis of discrete time varying "traffic." We next introduce combinatorial discrepancy and its adaptation to the analysis of time varying graphs.

5.3.1 Weighted Time Subsets

Assume that a time interval T of interest is given. A stream of T time-stamped and cd-labeled triplets $\mathbf{e} = \langle u, v, cd \rangle$ (that is, a shipment stream of contents labeled cd) can be viewed as a collection of weighted subsets of T, one for each $\mathbf{e}$. By considering only those nonzero entries in $M_{u,v}(t_i, cd)$ and letting $\mathrm{Vol}(\mathbf{e}, t_i) = M_{u,v}(t_i, cd) > 0$, we associated each pairing $\mathbf{e}$ with the multi-set $\{\mathrm{Vol}(\mathbf{e}, t_i)\}$ that encodes both the times and nonzero shipment volumes of cd contents from a foreign port u to a U.S. port v. Alternatively, each time t is weighted by the volume of cd units shipped at that time. We refer to the unweighted active times by $T(\mathbf{e}, t) = \{t_i{:}\mathrm{Vol}(\mathbf{e}, t_i){>}0\}$.

This view of a shipment stream (as a collection of weighted subsets of T, one per edge) enables the study of manifest data shipments as combinatorial set-systems. The discrepancy of such a system provides us with a novel mechanism to spot those port pairs (u, v) whose shipment traffic exhibits a pattern that is certainly out of the ordinary. Before we adapt the formulation of combinatorial discrepancy to the context of manifest data, we use the introduced notions to define a *Time Varying Manifest Shipment Graph* and its associated *Firing Rate Sequences*.

5.3.2 Manifest Data as a Time Varying Graph

The Manifest Shipment Graph MSG(T) is a time varying bipartite graph. One set of vertices consists of all port pairs $\langle u, v \rangle$ shipping contents during a time interval T, and the other set of vertices consists of all the content types appearing in the corresponding Contents Descriptors records. A pair $(\langle u, v\rangle, cd)$ is an edge e in the MSG(T) if at some time t in T, the foreign port u shipped contents of type cd to the U.S. port v. Each such edge e has associated the volume weighted sequence of time stamps $\{\mathrm{Vol}(e, t_i) = M_{u,v}(t_i, cd) > 0\}$. Each such weighted subset of T keeps track of the fact that at time t a shipment of type cd was sent between the pair of ports $\langle u, v\rangle$ with volume $\mathrm{Vol}(e, t_i) = M_{u,v}(t_i, cd) > 0$.

5.3.3 Cumulative Frequencies

For each edge e, we use the symbol $\|\mathrm{V}e, t\|$ to refer to the cumulative sum of Vol(e, t_i) for $t_i \leq t$ and we let Te,t denote the corresponding unweighted set of time stamps $\{t_i{:}\mathrm{Vol}(e, t_i) > 0\}$. The cardinality of T$e$,$t$ is then just the unweighted frequency of activity up to time t of the edge e.

5.3.4 Firing Rate Sequences

At each time t, each edge e = ((u, v), cd) has a "natural" weighted firing rate (velocity) equal to $\|\mathrm{V}(e, t)\|/t$. In the unweighted case, the firing rate becomes $|\mathrm{T}e, t|/t$. For a time interval T, the firing rate sequence f(e) of the edge $e = (\langle u, v\rangle, cd)$ is $f(e) = \{\|\mathrm{V}(e, t)\|/t$ for t in $T\}$. This can be naturally extended to any subset of edges, port pairs, or cd contents by summation.

Firing rate sequences provide a mechanism to pulse the traffic "behavior" of port pairs vs. contents, or subsets of port pairs (i.e., subgraphs). The central idea is to compare the firing rate sequences of subsets of (port pairs, contents) with the firing rate sequence of the overall shipment traffic in which they reside. This comparison is facilitated by the use of combinatorial discrepancy, which is introduced next.

5.4 Combinatorial Set System Discrepancy

The notation in this section follows that used in Abello et al. (2010). Given a collection of subsets $S_t = \{Te, t$: e is an edge of MSG(T)$\}$ and a two-coloring function Chi:$T \rightarrow \{-1, 1\}$, the *discrepancy* of the edge e at time t with respect to the coloring Chi is denoted by Chi(e, t) and is given by

$$\mathrm{Chi}(Te, t) = \sum \{\mathrm{Chi}(t_i) : t_i \in Te, t\}.$$

The discrepancy of a vertex with respect to a coloring function Chi is the sum of the discrepancies of the edges incident to it (similarly for any arbitrary but fixed subgraph *Z*). This Chi function keeps track through time of the "sign balance" of a time-ordered subsequence of elements of *T* with respect to the two-coloring Chi. The Chi_discrepancy of S_t is equal to the max{Chi(*Te*,*t*): *Te*,*t* in S_t}. *DISC*(S_t), the *discrepancy* of S_t, equals the minimum of the Chi_discrepancy(S_t) over all colorings Chi.

From basic results in combinatorial discrepancy (Chazelle 2000), it follows if t' is the maximum time when any edge is active, and $m_{t'}$ is the total number of edges active up to time t', the maximum discrepancy of our defined set systems is less than or equal to $\sqrt{2 \times t' \times \ln(2m_{t'})}$. We use this result to give to every edge *e* (and vertex *x*) at time *t*, a discrepancy-based weight Chi_Weightas follows:

$$\mathrm{Chi\_Weight}(e, t) = \left|\mathrm{Chi}(e, t) - \sqrt{2^* t'^* \ln(2m_{t'})}\right|$$
$$\mathrm{Chi\_Weight}(x, t) = \left|\mathrm{Chi}(x, t) - \sqrt{2^* t'^* \ln(2m_{t'})}\right|.$$

This Chi_Weight measures the difference between the Chi-value of an edge (or vertex) and the discrepancy upper bound.

5.4.1 Discrepant Edges and Vertices

An edge $e = (\langle u, v\rangle, cd)$ or vertex *x* is called *i-discrepant* if its discrepancy weight Chi_Weight(*e*, *t*) is *i*-standard deviations away from the mean of the distribution of the Chi-Weights of all the edges present up to time *t* of the form $(\langle u', v'\rangle, cd)$. Notice that this definition is for a fixed type of *cd* contents. It provides us with a mechanism to identify those port pairs $\langle u', v'\rangle$ whose behavior is "salient" with respect to shipments of a particular type. High absolute discrepancy is a good indicator for activity patterns substantially different from the activity pattern of the entire manifest data for a particular type of contents. Edges with quite different overall activities may have close Chi_Weight. This can be interpreted as a strong indicator that their activity patterns are "similar" even though one edge may be vastly more active than another. Those edges (vertices) with discrepancy weight close to zero tend to have activity patterns that are more difficult to spot, i.e., they are not very "salient" from the discrepancy point of view.

5.4.2 Coloring Function

Besides a random coloring, we used in our experiments a coloring that implicitly keeps track of long increasing or decreasing sequences of firing rates. This is done by comparing, at each time $t > 1$ of interest, the firing rate of the entire manifest shipment graph (a.k.a. the system) with the firing rate of an edge or vertex that is active at that time t. If both firing rates increase or decrease, we assign to that particular t the color value +1. If they disagree, we assign to that particular t the color value -1. We refer to this coloring as the ascend descend coloring. It encodes the level of firing rate agreement or disagreement that a particular edge or vertex has with respect to the firing rate of the entire system.

5.5 Fossil Visualization: Static Versus Dynamic Views

5.5.1 Static Views

Figures 5.4 and 5.5 are static views of the Manifest Shipment Graph. We use node link diagrams to depict the bipartite Manifest Shipment Graph described in Sect. 5.3. The two node types are differentiated by color. Green nodes represent port pairs and blue nodes represent contents. Edge attributes are encoded with edge color and edge thickness. The edge color encodes discrepancy weight with respect to ascend descend coloring or random coloring and it is modeled as a heat map, while the thickness of an edge encodes the firing rate.

High degree nodes represent highly active shipments. For instance, a blue node b with a high degree indicates that the content represented by b is being shipped by a high number of different port pairs. Similarly, a green node g with a high degree indicates that the port pair represented by g carries shipments with a large variety of different contents. As an example, in Fig. 5.4, node 030 representing the content <Business Industrial> and node 583092704 representing the port pair <Kaohsiung, China to Los Angeles, CA> are nodes with high shipment activity.

"Salient" edges (i.e., edges with maximum firing rate and maximum discrepancy weight) are represented as the thickest and reddest edges of the graph. This representation distinctly separates the "salient" edges from the remaining edges. "Salient" edges are those that have been highly active up to time t and are currently those edges that deviate the most from the overall behavior of the entire system.

5.5.2 Dynamic Views

Even though color and thickness convey useful static information, about vertices and edges, they are limited in capturing time varying behavior. In order to encode time varying patterns, we need a mechanism that keeps track of "salient" nodes or

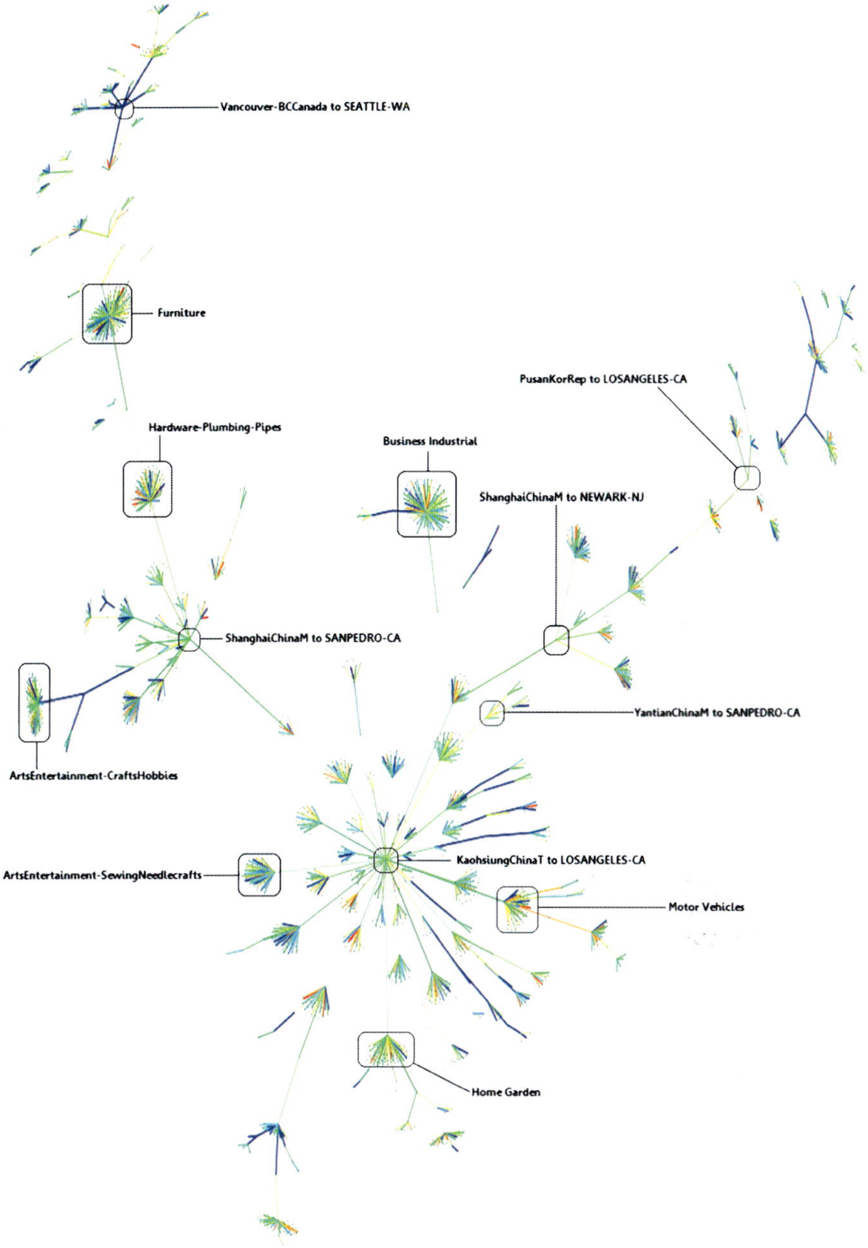

Fig. 5.4 A visualization of the cumulative maximum spanning forest with ascend descend coloring of manifest data from January 30, 2009, up to February 28, 2009. Neighborhoods containing port pair and content nodes with high degree are labeled

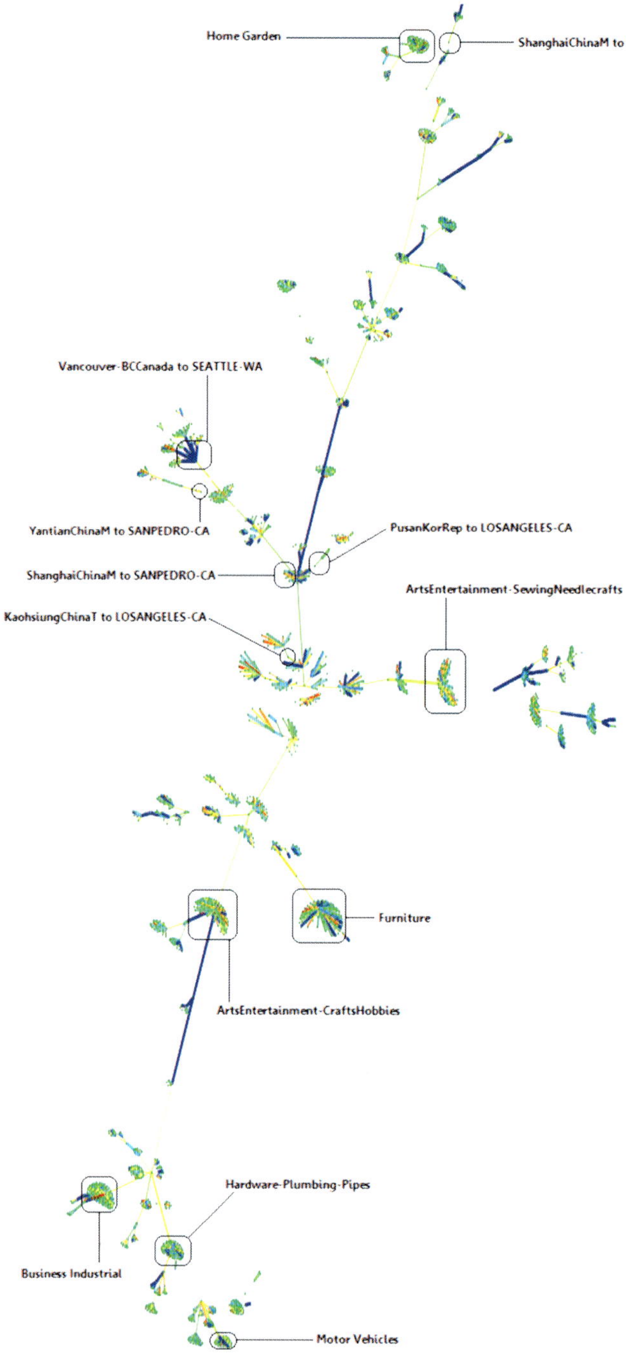

Fig. 5.5 A visualization of the cumulative maximum spanning forest with random coloring of manifest data from January 30, 2009, up to February 28, 2009. Neighborhoods containing the same nodes from Fig. 5.4 are labeled

edges as they occur in the input stream. Such a mechanism must maintain a global sketch of the entire graph together with a "few" local markers of temporal notoriety. Temporal notoriety is maintained by the discrepancy weight with respect to the random coloring and ascend descend coloring. Discrepancy with respect to random coloring targets those edges whose activity sequence exhibits some clearly discernible pattern (i.e., they are not random). On the other hand, discrepancy with respect to the ascend descend coloring detects those edges with substantial time periods of bursty behavior. Their combination marks at each time t those few edges whose time series activity is "peculiar" in the context in which they occur in the data stream. Collecting these "peculiar" edges through time provides an automatic summary (i.e., a small subgraph) of the evolving input stream. We call this automated summary a *fossil*. To maintain a global mental image of the evolving edge stream, we maintained a maximum discrepancy spanning forest (MDSF) into which the inclusion of an edge is governed by its discrepancy weight with respect to an a priori chosen coloring scheme (Abello et al. 2010). The MDSF provides a mental map of the data, and by restricting the fossil edges to be part of the MDSF, we obtained a fossil subgraph that records the most temporal "salient" edges. In summary, by computing a time varying maximum discrepant spanning forest of the manifest shipment graph, we extracted those triples <foreign port, U.S. port, ContentsDescriptor> whose shipment activity exhibits unusual patterns of bursty behavior. Notice that what we call a "fossil" is more than just a selection of those edges with high discrepancy at a particular time t. That is, an edge becomes part of the fossil only if it is selected as a highly discrepant edge a "high" number of times. It is truly a record of time varying shipment activity patterns. That is why we refer to it as a "fossil." As seen in Figs. 5.6 and 5.7, a subset of all edges in the fossils are highlighted and marked according to the respective coloring schemes (See Appendix 2 for sample edges included in the fossil and Appendix 3 for a complete list of "salient" edges selected by both coloring schemes through time.).

5.6 Conclusions and Future Work

This chapter presented several novel techniques to extract from manifest data "unusual" time varying shipment patterns. The proposed techniques are based on content vector representations of each port pair. From this collection of vectors, a variety of weighted similarity graphs were derived. The edge weights were functions of either the dot product of the associated content vectors or their Hellinger distance (after suitable normalization). Clustering the port pair's content vectors, according to their similarity, produced a small collection of shipment patterns that can be succinctly described. Classifying different days of shipment patterns via their Hellinger distances agreed with the popular belief that "Same weekdays have similar traffic patterns" or equivalently "All Mondays are the same, all Tuesdays are the same, etc." On the other hand, at a quite refined level of granularity, we are able to extract those triples <*u*, *v*, *contents*> that exhibited "unusual" or "salient" time varying shipment patterns.

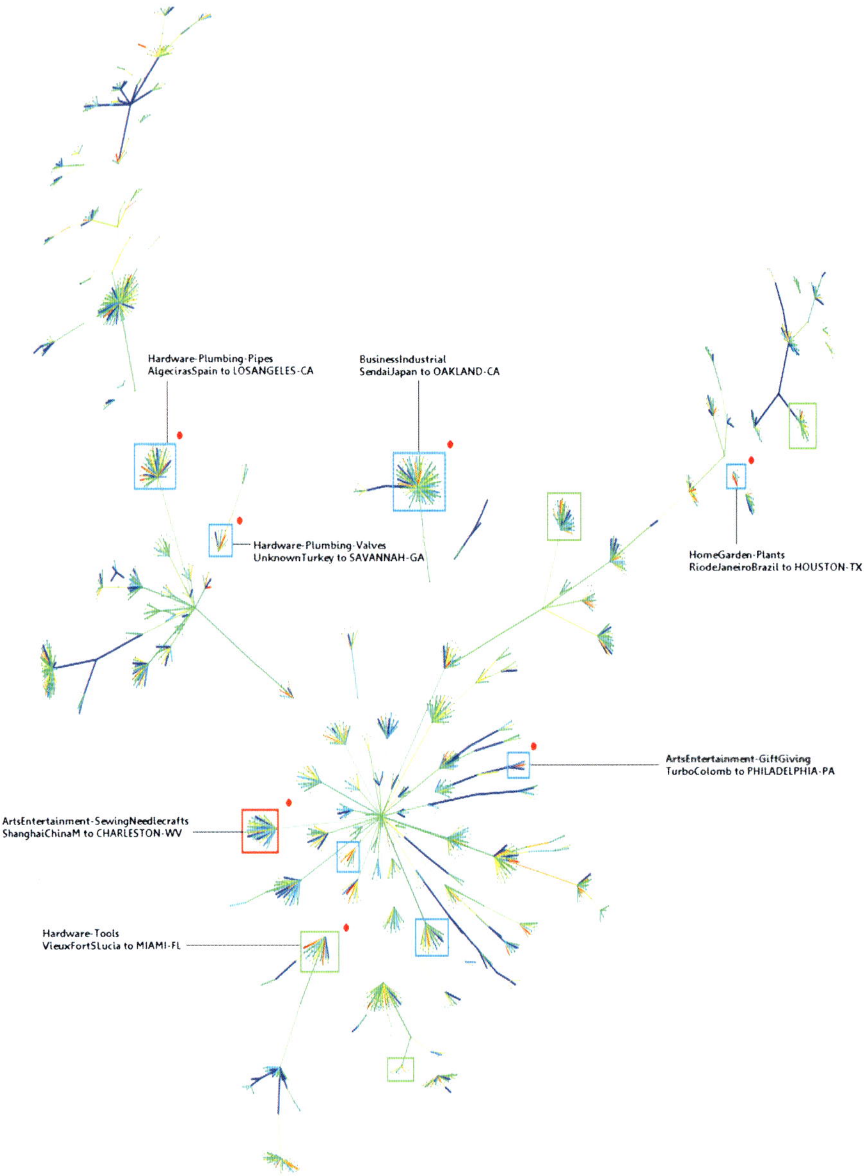

Fig. 5.6 Neighborhoods bordered by *boxes* indicate that at least one edge appeared in the fossil. *Colors of the bordering boxes* correspond to the selected edge frequency in the fossil with *red* being highest, *blue* being lowest. *Labeled boxes* marked by a *red dot* indicate neighborhoods that contain an edge that was selected by both coloring schemes

This was achieved by using an adaptation of combinatorial set system discrepancy to the context of time varying graphs (Abello et al. 2010). The central idea is to associate with each triple a weighted time sequence and to track the weighted cumulative behavior of its associated firing rate and discrepancy sequences. Those triples whose

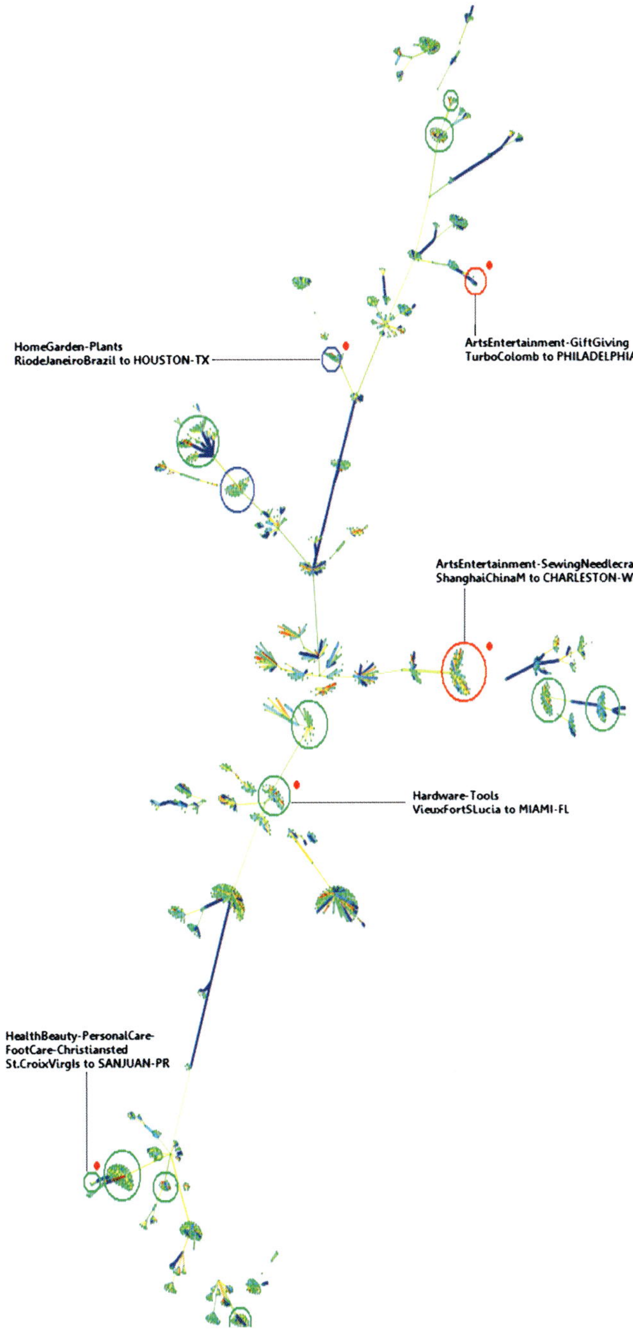

Fig. 5.7 Neighborhoods bordered by *circles* indicate that at least one edge appeared in the fossil. *Colors of the bordering circles* correspond to the selected edge frequency in the fossil with *red* being highest and *blue* being lowest. *Labeled circles* marked by a *red dot* indicate neighborhoods that contain at least one edge that was selected by both coloring schemes

discrepancy and firing rate were above a certain number of standard deviations from the mean discrepancy weight of the entire collection of shipments were selected as the most "salient" triples, i.e., the selected triples were highly discrepant and exhibited relatively long intervals of bursting behavior. This collection of selected triples (a fossil) constituted a succinct time varying representation of the most salient port pairs for particular content types. To keep track of the context in which fossil triples occur, we maintained a time varying MDSF in which the fossil triples were embedded. The proposed techniques are applicable to a variety of time varying and edge-labeled social network graphs such as those extracted from Twitter data.

In general, it will be interesting to elucidate the complexity of computing the MDSF of a time varying graph when different edge instances occur in a completely distributed setting. From the graph drawing arena, the central problem is how to compute dynamically the coordinates of the vertices of a time evolving maximum spanning forest so that a mental map of the forest evolution is minimally disrupted. At the user interface level, what are the fundamental interaction mechanisms that are useful aids for the navigation, exploration, and summarization of time varying data? Systems like Gephi (Bastion et al. 2009) and GraphView (Abello et al. 2010) have been very useful to us in some respect, but useful interaction for clear productive exploration of this type of time varying data is certainly in its infancy.

Acknowledgments We thank Tsvetan Asamov, Jerry Chen, and Nishchal Devanur for collaborations at the initial stages of this project. We acknowledge the support provided by DHS Agreement Number 2008-DN-077-ARSI012-04, DIMACS Special Focus on Algorithmic Foundations of the Internet, NSF Grant #CNS-0721113, and mgvis.com.

Appendix 1: Sample Records of Original Data

Human Readable version

```
Bill Info:
        CARRIER-CODE=          CSVV
        VESSEL-COUNTRY-CODE=            US
        VESSEL-NAME=           BRAZILIAN REEFER
        VESSEL-NUMBER=         32N
        DP-OF-UNLADING=        2704
        EST-ARRIVL-DATE=       70215
        BILL-OF-LADING-NBR=    CSVVCFE078240
        FOREIGN-PORT-LADING=            33797
        MANIFEST-QTY=          2400
        MANIFEST-UNIT=         PCS
        WEIGHT=   18000
        WEIGHT-UNIT=           KG
        MEASUREMENT-UNIT=      CM
        DELETE-INDICATOR=      D
        PLACE-RECEIPT=         VALPARAISO, CHILE
        VESSEL-CODE=           8300377
        MODE-TRANSPORT=        10
        MANIFEST-NBR=          740
        ACT-ARRL-DATE=         20070215

 --------------------------------------------------
```

```
Container Info:
      CONTAINER-1=          GESU9161211
      SEAL-1-1=  K171013
      EQ-DESC-CODE-1=       30
      CONT-LENGTH-1=        4000
      CONT-HEIGHT-1=        906
      CONT-WIDTH-1=         800
      CONT-TYPE-1=          4530
      LOAD-STATUS-1=        L
      TYPE-SERVICE-1=       PP

---------------------------------------------------
ConSignee Info:
      CONSIGNEE-NAME=       FUNNY FRUIT SALES CO.
      CONSIGNEE-ADDR-1=     Address Details        City      State 99999-0000
      CONSIGNEE-ADDR-2=     CA 93720 PHO:(xxx-xxx-xxxx)
      CONSIGNEE-ADDR-3=     CONTACT: xxxx xxxxxx

---------------------------------------------------
Nofiy Info:
      NOTIFY-NAME=          J&K NOFRESH LLC.
      NOTIFY-ADDR-1=        Address Details City      State 99999-0000
      NOTIFY-ADDR-2=        CITY CA 99999-0000 CONTACT
      NOTIFY-ADDR-3=        PERSON: yyyyy yyyyyyyy  PH:yyy-yyy-
      NOTIFY-ADDR-4=        8770 FAX: yyy-yyy-yyyy

---------------------------------------------------
Other Contain Info:
      CONTAINER-NUMBER=  GESU9161211
      PIECE_COUNT-1=        2400
      DESCRIPTION-1=        AS DESCRIBED FRESH PLUMS CARTON
      DESCRIPTION-2=        VARIETY:
      DESCRIPTION-3=        FRIAR 7,5KB.

      CONTAINER-NUMBER=  GESU9161211
      DESCRIPTION-1=        E-156324 REF:037-07
      DESCRIPTION-2=        REFRIGERATED CARGO
      DESCRIPTION-3=        20 PALL

      CONTAINER-NUMBER=  GESU9161211
      DESCRIPTION-1=        USDA INSPECTED

---------------------------------------------------
MRKS-NBRS Info:
      CONTAINER-NUMBER=  GESU9161211
      DETAILS-1= TRINIDAD
      DETAILS-2= GESU916121-1
      DETAILS-3= SEAL:K171013
      DETAILS-4= LCL/LCL
```

Corresponding Original Text Data

```
1CSVVUSBRAZILIAN REEFER      0032N2704070215CSVVCFE078240   3379700000002400PCS
00000018000KG00000000000CMDVALPARAISO, CHILE         830037700100007400
20070215
1GESU9161211  K171013                 3004000000000906000008004530LPP
FUNY FRUIT SALES CO.          Address Details City      State 9999-0000   93720 PHO:(xxx-xxx-xxxx)
CONTACT:xxxxx xxxxxx
4J&K NOFRESH LLC.             Address Details           City      State 9999-0000    CONTACT
PERSON: yyyyy yyyyyyyy  PH:yyy-yyy-yyyy    FAX:xxx-xxx-xxxx
```

Appendix 2: Sample Fossil Information as Seen in Figs. 5.6 and 5.7

Recall that an edge is selected to be in the i-fossil at time t if its weight is at least i-standard deviations away from the overall data discrepancy at that time. The following are samples of edges in the i-fossil, where i is the maximum number of standard deviations at time t. Our fossil frequency calculations ignore timestamp 1 (denoted as FOS_FREQ in the tables below).

Ascend Descend Coloring Fossil

EDGE_ID	DISC_WGT	FOS_FREQ	ALL_TIMES	ACTUAL_ID
"February 16th, 2009 (Monday)"				
393-588402811	24.1660748	3	"3,17,18"	MotorVehicles-AutomotiveParts-AirIntakeParts-KobeJapan→OAKLAND-CA
"February 18th, 2009 (Wednesday)"				
238-248675201	26.3931427	3	"2,13,20"	Hardware-Tools-VieuxFortSLucia→MIAMI-FL
392-243115203	27.7802982	3	"1,5,12,22"	MotorVehicles-AutomotiveParts-AutomotiveFuelSystems-ProvidencialesTurkIs→PORTEVERGLADES-FL
"February 21st, 2009 (Saturday)"				
015-570351409	28.4691391	3	"2,9,23"	ArtsEntertainment-CraftsHobbies-SewingNeedlecrafts-ShanghaiChinaM→CHARLESTON-WV
"February 22nd, 2009 (Sunday)"				
015-570351409	28.4691391	4	"2,9,23,24"	ArtsEntertainment-CraftsHobbies-SewingNeedlecrafts-ShanghaiChinaM→CHARLESTON-WV
045-307645201	28.4691391	3	"2,23,24"	BabyToddler-NursingFeeding-PuertoCabelloVenez→MIAMI-FL
072-602672704	28.4691391	2	"23,24"	ClothingAccessories-ClothingAccessories-Veils-SydneyAustral→LOSANGELES-CA
227-470312704	28.4691391	2	"23,24"	Hardware-Plumbing-Pipes-AlgecirasSpain→LOSANGELES-CA
"February 23rd, 2009 (Monday)"				
030-588972811	29.7280884	2	"3,25"	BusinessIndustrial-SendaiJapan→OAKLAND-CA
397-301401103	29.7280884	2	"4,25"	MotorVehicles-AutomotiveParts-OilCirculation-SantaMartaColomb→WILMINGTON-DE
"February 24th, 2009 (Tuesday)"				
015-205131902	30.4009418	3	"4,11,26"	ArtsEntertainment-CraftsHobbies-SewingNeedlecrafts-QuatemaGuatmal→GULFPORT-MS
231-489471703	30.4009418	2	"2,26"	Hardware-Plumbing-Valves-UnknownTurkey→SAVANNAH-GA
"February 25th, 2009 (Wednesday)"				
016-301451101	31.0294342	2	"14,27"	ArtsEntertainment-GiftGiving-TurboColomb→PHILADELPHIA-PA
"February 26th, 2009 (Thursday)"				
238-248655201	31.6441116	2	"20,28"	Hardware-Tools-CastriesSLucia→MIAMI-FL
"February 28th, 2009 (Saturday)"				
336-351715301	32.8434029	2	"1,13,30"	HomeGarden-Plants-RiodeJaneiroBrazil→HOUSTON-TX

Random Coloring Fossil

EDGE_ID	DISC_WGT	FOS_FREQ	ALL_TIMES	ACTUAL_ID
"February 13th, 2009 (Friday)"				
002-538271703	21.798954	3	"6,14,15"	Animals-PetSupplies-ChittagongBngldsh→SAVANNAH-GA
002-582015301	21.798954	3	"8,14,15"	Animals-PetSupplies-HongKongHgKong→HOUSTON-TX
016-301451101	21.798954	3	"6,14,15"	ArtsEntertainment-GiftGiving-TurboColomb→PHILADELPHIA-PA
030-791135301	21.798954	3	"5,14,15"	BusinessIndustrial-PortNatalRepSAF→HOUSTON-TX
254-570350403	21.798954	3	"7,14,15"	Hardware-Tools-WireCableTools-ShanghaiChinaM→WORCESTER-MA
278-911154909	21.798954	3	"2,14,15"	HealthBeauty-PersonalCare-FootCare-Christiansted-St.CroixVirgIs→SANJUAN-PR
289-588571001	21.798954	3	"5,14,15"	HomeGarden-Decor-NagoyaKoJapan→NEWYORK-NY
290-580235203	21.798954	3	"5,14,15"	HomeGarden-Decor-DoorsWindows-PusanKorRep→PORTEVERGLADES-FL
315-351701401	21.798954	3	"7,14,15"	HomeGarden-KitchenDining-Appliances-SuapeBrazil→PORTSMOUTH-VA
374-588955203	21.798954	3	"8,14,15"	MotorVehicles-YokohamaJapan→PORTEVERGLADES-FL
416-559762904	21.798954	3	"6,14,15"	OfficeSupplies-PaperHandling-SingaporeSingapr→PORTLAND-OR
462-241281801	21.798954	3	"5,14,15"	ToysGames-PortBustamanteJamaica→TAMPA-FL
238-248675201	26.3931427	3	"2,13,20"	Hardware-Tools-VieuxFortSLucia→MIAMI-FL
"February 21st, 2009 (Saturday)"				
015-570351409	28.4691391	3	"2,9,23"	ArtsEntertainment-CraftsHobbies-SewingNeedlecrafts-ShanghaiChinaM→CHARLESTON-WV
"February 22nd, 2009 (Sunday)"				
015-570351409	28.4691391	4	"2,9,23,24"	ArtsEntertainment-CraftsHobbies-SewingNeedlecrafts-ShanghaiChinaM→CHARLESTON-WV
"February 25th, 2009 (Wednesday)"				
016-301451101	31.0294342	4	"6,14,15,27"	ArtsEntertainment-GiftGiving-TurboColomb→PHILADELPHIA-PA
"February 26th, 2009 (Thursday)"				
058-351731401	31.6441116	2	"7,28"	ClothingAccessories-Clothing-Tops-RioGrandeBrazil→PORTSMOUTH-VA
"February 28th, 2009 (Saturday)"				
336-351715301	32.8434029	2	"1,13,30"	HomeGarden-Plants-RiodeJaneiroBrazil→HOUSTON-TX

Appendix 3: Salient Edges Selected by Both Coloring Schemes

The following are examples of edges that are included in the fossil with respect to both random coloring and ascend descend coloring.

570672811 (UnknownChinaM to Oakland-CA) - 009 (ArtsEntertainment-CraftsHobbies)
570351409 (ShanghaiChinaM to Charleston-WV) - 015 (ArtsEntertainment-SewingNeedleCrafts)
205131902 (QuatemaGuatmal to GulfPort-MS) - 015 (ArtsEntertainment-SewingNeedleCrafts)
301451101 (TurboColomb to Philadelphia-PA) - 016 (ArtsEntertainment-GiftGiving)
588972811 (SendaiJapan to Oakland-CA) - 030 (BusinessIndustrial)
331314601 (GuayaquilEcuador to Newark-NJ) - 030 (BusinessIndustrial)
570511001 (XingangChinaM to NewYork-NY) - 058 (ClothingAccessories-Tops)
247215201 (BocaChicaDomRep to Miami-FL) - 074 (ClothingAccessories-Footwear)
552241001(ThanhPhoHoChiMinhVietnam to NewYork-NY) - 171 (Furniture-BedroomFurniture)
333632501 (PaitaPeru to SanDiego-CA) - 189 (Food-DessertWine)
307641803 (PuertoCabelloVenez to JacksonVille-FL) - 216 (Hardware-Plumbing)
470312704 (AlgecirasSpain to LosAngeles-CA) - 227 (Hardware-Plumbing-Pipes)
489471703 (UnknownTurkey to Savannah-GA) - 231 (Hardware-Plumbing-Valves)
248675201 (VieuxFortSLucia to Miami-FL) - 238 (Hardware-Tools)
911154909 (Christiansted-St.CroixVirgIs to SanJuan-PR) - 278 (HealthBeauty-FootCare)
935013002 (GuamIslandGuam to Tacoma-WA) - 285 (HomeGarden)
223131102 (PuertoLimonCRica to Chester-PA) - 330 (HomeGarden-Lighting)
351715301 (RiodeJaneiroBrazil to Houston-TX) - 336 (HomeGarden-Plants)
272015201 (BocaChicaDomRep to Miami-FL) - 364 (Media-HardRockMetal)
583093801 (KaohsiungChinaT to Wyandotte-MI) - 382 (MotorVehicles-AutomotiveLockingSystems)
236725203 (PortAndrosBahamas to PortEverglades-FL) - 392 (MotorVehicles-AutomotiveFuelSystems)
201522101 (CayoArcosTerminalMexico to PortArthur-TX) - 397 (MotorVehicles-OilCirculation)
535501001 (KarachiHarborPakistn to NewYork-NY) - 406 (OfficeSupplies)

References

Abello J, Eliassi-Rad T, Devanur N (2010) Detecting novel discrepancies in communications networks. In: International conference on data mining, ICDM Dec 2010, Sydney, Australia, pp 8–17

ASFOI.txt (2009) Department of Homeland Security, DyDan Center, DIMACS, Rutgers University

Bastion M, Heymann S, Jacomy M (2009) Gephi: an open source software for exploring and manipulating networks. In: Proceedings of the third international conference on weblogs and social media, May 2009

Chazelle B (2000) The discrepancy method: randomness and complexity. Cambridge University Press, New York

Lin J (1991) Divergence measures based on the Shannon entropy. IEEE Trans Inform Theory 37(1):145–151

Chapter 6
Achieving Realistic Levels of Defensive Hedging Based on Non-monotonic and Multi-attribute Terrorist Utility Functions

Vicki Marion Bier, Jaime Marie Bonorato, and Chen Wang

Abstract This chapter addresses the problem of allocating limited resources to defend a set of targets. When there is uncertainty about which targets the terrorists are most likely to attack, decision makers are likely to insist on some degree of "hedging" (defending targets with only moderate value). The work discussed in this chapter uses game theory to find the optimal strategy for the defender and shows that non-monotonic attacker objective functions do typically yield greater hedging.

6.1 Introduction

In the years following September 11, 2001, funding for homeland security grew dramatically. For example, in the first 3 years following September 11, 2001, federal expenditures allocated to terrorism prevention and response increased nearly 1,000% (from \$1.2 billion to \$13.1 billion) (Ripley 2004).

Prior to the terrorist attacks of September 11, 2001, the criteria for deciding how to allocate defensive resources depended mostly on population statistics (Davis 1998). After September 11, 2001, in response to an increased awareness of terrorism, more complex decision methodologies were developed to meet the need for more robust strategies for resource allocation (Walker 2002). In particular, allocations based on risk began to appear, and population statistics were no longer the only deciding factor. However, many funds (e.g., funding under the State Homeland Security Program and the Citizen Corps Program) were initially distributed based on an approximate 60/40 split, with only about 60% of funds distributed according to risk methodologies;

V.M. Bier (✉)
Department of Industrial and Systems Engineering, University of Wisconsin-Madison, 1513 University Avenue, Madison, WI 53706, USA
e-mail: bier@engr.wisc.edu

J.M. Bonorato • C. Wang
University of Wisconsin-Madison, Madison, WI, USA

J.W. Herrmann (ed.), *Handbook of Operations Research for Homeland Security*, International Series in Operations Research & Management Science 183,
DOI 10.1007/978-1-4614-5278-2_6,

the remaining 40% were statutorily disbursed, such that each state received an equal share. By contrast, other grant programs, such as the Urban Areas Security Initiative and the Metropolitan Medical Response System, were available only to regions determined to be eligible based on specific criteria (http://www.ojp.usdoj.gov/odp/grants_program.htm). Moreover, risk-assessment methodologies themselves were still based primarily on population measures (Brunet 2006).

Decision methodologies continued to evolve in fiscal years (FY) 2003 and 2004, when risk estimation began to include factors other than population measures alone. The factors considered at this time included threat (both credible threats identified by the intelligence community and the results of field investigations by the Federal Bureau of Investigation and Immigration and Customs Enforcement), the presence of critical infrastructure, and population density. However, these factors were still aggregated in an additive rather than a multiplicative manner, and probabilities were not included (Masse et al. 2007).

A dramatic reform was first seen in FY 2006, due in part to the influence of Michael Chertoff as the new Secretary of DHS. Chertoff had promised to adopt a risk-based decision-making approach (Asaba 2006), so, for the first time, funding formulas included estimates of the probabilities of certain events. In particular, risk was now calculated by multiplying threat (the likelihood of an attack occurring), vulnerability, and consequence (where the product of vulnerability and consequence reflected both relative exposure and the expected impact of an attack). In addition, threat estimates from the intelligence community and inherent risks associated with different geographic areas (such as international borders) were now considered (Reese 2006). DHS also began to include an effectiveness assessment, requiring urban areas applying for grants under the Urban Areas Security Initiative to submit information on the anticipated effectiveness of their proposed solutions to homeland security needs.

In FY 2007 and 2008, DHS's methodologies continued to improve—in particular, by integrating geographic and asset-based assessments (as opposed to considering them separately), basing threat on assessments by the intelligence community (with each urban area assigned to one of four tiers), and better defining the consequences of attacks. However, DHS still considered all areas equally vulnerable to attack, because of the ease of mobility throughout the country; therefore, each area was assigned a vulnerability score of one (GAO 2008).

Although the allocation process has advanced since 2001, further improvement is still needed to take into account two fundamental factors—the strategic nature of terrorist actions and defender uncertainty about terrorist motivations and goals. For example, the National Research Council (2008) recommended that "to assess the probabilities of terrorist decisions, DHS should use elicitation techniques and decision-oriented models that explicitly recognize terrorists as intelligent adversaries who observe U.S. defensive preparations and seek to maximize achievement of their own objectives." Parnell et al. (2008), likewise, argued that terrorists are "goal-oriented, resourceful adversaries, who will, given the constraints they perceive, select the best agent and target to achieve their objectives." With respect to uncertainty, a variety of game-theoretic models in the face of defender uncertainty have been studied and applied; see, for example, the reviews by Sandler and Siqueira (2009) and Guikema (2009). However, these models have not yet been widely used in applications.

Bier et al. (2007) proposed a game-theoretic method for identifying attacker and defender equilibrium strategies in a sequential game, where the defender plays first by allocating defensive resources among possible terrorist targets, followed by the attacker deciding on which target to attack, in light of any defensive investments. Their model assumes that the defender does not have full knowledge about the attacker's preferences; thus, the defender's objective is to minimize the expected loss from any successful (or attempted) attack(s), in light of defender uncertainty. This model should, in principle, provide a more rational and rigorous basis for defenses against uncertain and adaptive adversaries than earlier game-theoretic models that do not consider defender uncertainty. However, applications of this model (e.g., Bier et al. 2008) still yield relatively little defensive "hedging" (i.e., little investment in targets of only moderate value to the defender), even when the defender is assumed to be quite uncertain about attacker preferences.

Therefore, in this chapter, we consider additional applications and modifications of the model by Bier et al. (2007) to increase its usefulness for optimization of defensive allocations against terrorism. We are especially interested in examining the model's ability to yield realistic levels of defensive hedging in resource allocation. In particular, in the real world, some level of hedging will typically be desirable, since defenders generally won't know an attacker's goals perfectly. However, most game-theoretic models to date (even ones that explicitly consider defender uncertainty) often yield defensive allocations in which only the top few targets are protected, regardless of the extent of defender uncertainty (at least when defensive investments are not highly cost-effective), which seems unrealistic. At the same time, in order to allocate funds efficiently, we would still like to avoid spending too much money on targets that are relatively unlikely to be attacked (excessive hedging). If game-theoretic models are to become sufficiently mature for use in practice, models that result in realistic levels of hedging will be needed.

In an attempt to achieve more realistic levels of defensive hedging, we extend the model developed by Bier et al. (2007) to allow for both non-monotonic terrorist objective functions and multi-attribute terrorist objective functions. First, we consider a simple case study involving allocation of a limited defensive budget to ten major US cities. We assume that both the defender and the attacker care about fatalities, but allow the attacker to have a non-monotonic utility function over fatalities, assuming that moderate numbers of fatalities resulting from an attack would be preferred by the attacker over either too many or too few fatalities. This might be the case, for example, if large numbers of fatalities would lead to reduced support for the terrorist's cause, or massive US retaliation.

Then, we consider a more complex case study (also with a constrained budget), based on the balanced-scorecard model of target attractiveness developed by Beitel et al. (2004). They consider multiple attacker attributes (such as the logistical difficulty and resource requirements of particular attack strategies) in addition to the consequences of attacks to the defender and provide estimates reflecting assumed Al-Qaeda valuations of those attributes for targets such as US buildings, US corporate interests, and various transportation assets. We use these estimates from Beitel et al. as attribute values in the attacker's hypothesized multi-attribute utility function and treat the weights on those attributes as random variables (to reflect defender uncertainty about attacker preferences). We then explore the extent to which each of these extensions helps to achieve reasonable levels of hedging in recommended defensive investments.

6.2 Basic Game-Theoretic Model

The basic model of Bier et al. (2007) assumes that the defender's objective is to minimize the total expected loss from terrorist attacks, as given by

$$\min_{c_1,\ldots,c_n} \sum_{i=1}^{n} h_i(c_1,\ldots,c_n)p(c_i)v_i.$$

Subject to the budget constraint:

$$\sum_{i=1}^{n} c_i \leq B$$

where:

n = number of targets

c_i = defender's resource allocation to target i

B = defender's total budget

vi = defender's valuation of target i

$h_i(c_1,\ldots,c_n)$ = probability of an attack on target i

$p(c_i)$ = success probability of an attack on target i, as a function of the resources allocated to protection of target i

The attacker observes the defender's resource allocations c_i and then chooses the target with the highest payoff in light of any defensive investments,

$$\max_i p(c_i)U_i$$

where U_i is the attacker's utility of target i. This determines $h_i(c_1,\ldots,c_n)$. Note that since the defender is assumed to be uncertain about the attacker's preferences, the attacker target valuations U_i are modeled as random variables.

As in Bier et al. (2008), the success probability of an attack on target i is assumed to be given by $p(c_i) = e^{-\lambda c_i}$, so that λ is a measure of the cost-effectiveness of defensive investment. Cost-effectiveness in this model can be thought of as a measure of risk reduction per dollar spent. For example, at a value of $\lambda = 0.02$, if the c_i are measured in millions of dollars, then every million dollars of defensive investment would reduce the success probability of an attack by about 2%.

In the following sections, we apply this model with two different assumptions regarding the attacker's target valuations U_i. In the first case, we assume that the utilities U_i are non-monotonic functions of the target attribute values; next, we model the utilities U_i by a multi-attribute utility function with uncertain attribute weights. Sensitivity analysis is conducted for both cases to see how the extent of hedging depends on parameters such as the cost-effectiveness of defensive investment and the extent of defender uncertainty.

6.3 Non-monotonic Terrorist Utility

Here, we examine the effects of a non-monotonic attacker objective function on optimal defender resource allocations in an attempt to increase the realism and flexibility of the model outlined above and explore whether such model formulations give rise to significantly greater defender hedging at optimality. In this case, the attacker objective is still to impose damage on the defender; however, the model recognizes that attackers may not want to inflict as much damage as possible. For example, attackers may not want to "set the bar too high" such that they cannot follow up with an equal or greater impact in future attacks. Likewise, extremely damaging attacks may yield reduced benefits to the attacker due to adverse outcomes such as negative publicity, retaliation by the defender, or revulsion among those who had previously supported the attacker's cause.

The data used in this section come from Willis (2005), who provide estimates for three types of damage (property losses, fatalities, and total injuries) from attacks on different urban areas in the USA. For simplicity, we restrict our analysis to the case where the defender and the attacker both care only about fatalities. In particular, the defender's disutility is assumed to be proportional to (and thus strictly increasing in) the number of fatalities. By contrast, the attacker is assumed to have a non-monotonic utility function over fatalities, preferring a moderate number of fatalities over either extremely small numbers of fatalities (which may not have the desired impact) or extremely large numbers of fatalities (which may lead to reduced support or massive US retaliation).

For reasons of computational tractability, we restrict our attention to the ten urban areas of the USA that are estimated to have the highest expected terrorism fatalities in Willis (2005): New York City (NYC); Chicago; the Washington, D.C., area (including parts of Maryland, Virginia, and West Virginia); San Francisco; Los Angeles and Long Beach; the Boston area (including parts of New Hampshire); Houston; the Philadelphia area (including parts of New Jersey); Newark; and the Seattle area (including Bellevue and Everett). The expected annual fatalities from terrorist attacks for the top ten US urban areas are shown in Table 6.1.

The non-monotonic attacker utility U_i for target i is described below as a piecewise function of the expected fatalities y_i, and graphed in Fig. 6.1:

$$U_i = 110 - \frac{300}{y_i} \quad \text{for} \quad y_i < 30 \quad \text{and} \quad U_i = 110 - \frac{y_i}{3} \quad \text{for} \quad y_i \geq 30.$$

Table 6.1 Expected annual terrorism fatalities (taken from Willis 2005)

Urban area	Expected fatalities
NYC	304
Chicago	54
Washington, D.C.	29
San Francisco	24
Los Angeles	17
Boston	12
Houston	9
Philadelphia	9
Newark	4
Seattle	4

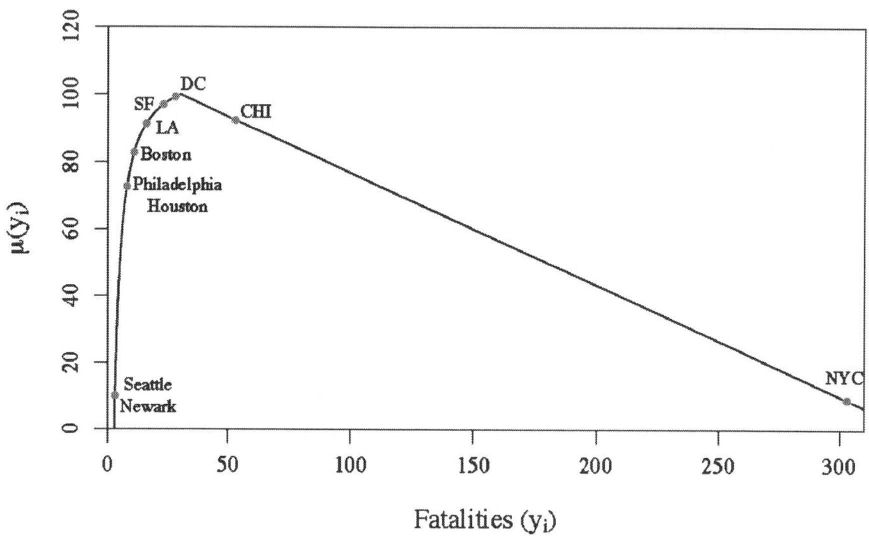

Fig. 6.1 Non-monotonic attacker preferences over expected terrorism fatalities

As shown in Fig. 6.1, Washington, D.C., is assumed to be the most attractive city to the attacker, even though it is only the third highest in terms of expected fatalities (an order of magnitude less than NYC). By contrast, NYC becomes one of the least attractive targets, due to the extremely large number of expected fatalities associated with an attack there.

Figure 6.2 shows the optimal budget allocation as a function of the cost-effectiveness of defensive investments, for both a monotonic attacker utility function (assumed to be proportional to the number of fatalities) and the above non-monotonic attacker utility function. As expected, the non-monotonic attacker objective function (Fig. 6.2b) does lead to dramatically greater defensive investment in cities of only moderate value to the defender than the monotonic objective function (Fig. 6.2a), especially at low levels of cost-effectiveness. Since mid-level

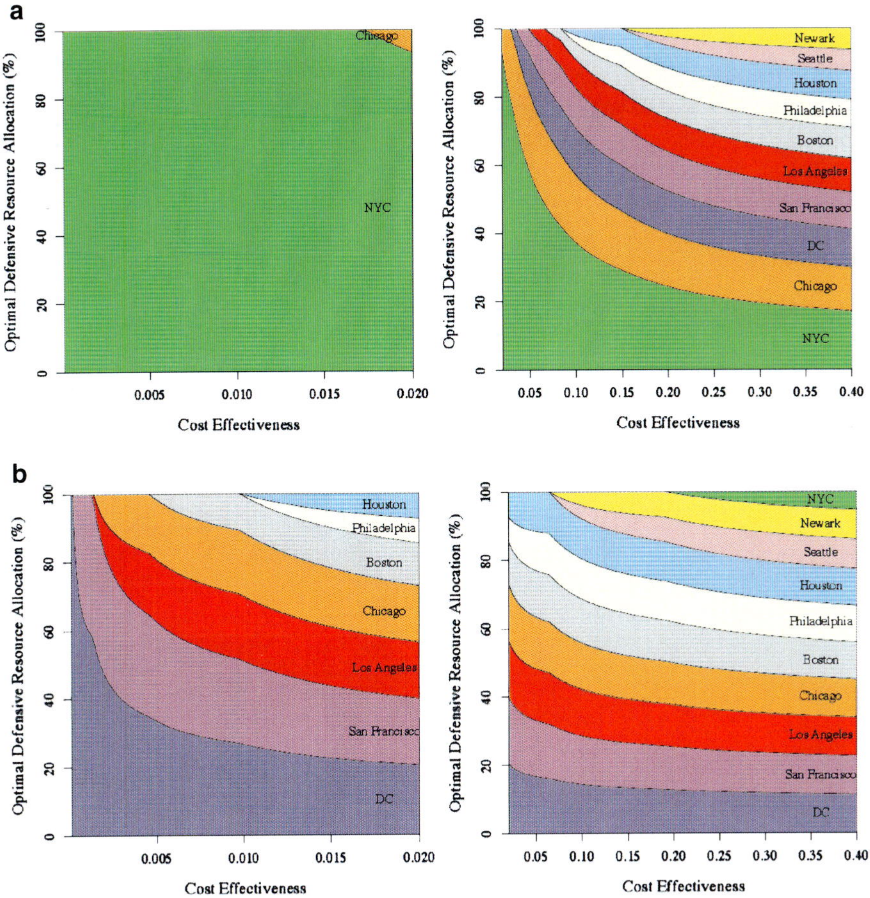

Fig. 6.2 Optimal defensive resource allocations to the ten major US urban areas, monotonic versus non-monotonic attacker preferences. (**a**) Monotonic attacker utility function. (**b**) Piecewise non-monotonic attacker utility function

cities such as Washington, D.C., San Francisco, and Los Angeles are assumed to be more attractive to the attacker in the non-monotonic case, those targets receive the most funding in that case, as opposed to high-valued cities such as NYC in the monotonic case. In fact, in the non-monotonic case, NYC gets no defensive investment at all when the cost-effectiveness of defensive investment is low. This is because NYC is believed to be unlikely to be attacked for this particular choice of non-monotonic attacker objective function, and thus does not rise to a level of risk that would justify funding when defensive investment is not highly cost-effective.

Of course, in reality, even if the attacker prefers a medium-sized attack to a larger attack, the attacker could still choose to attack a high-value target such as NYC, and just launch a smaller attack there. Therefore, the fact that our optimal resource allocation does not allocate any money to NYC at low levels of cost-effectiveness in the non-monotonic case is an unrealistic feature of our model. The results of the model should thus not be taken to suggest that, at low cost-effectiveness, we should not defend NYC or other high-valued cities. However, the results are sufficient to indicate that, if we believe the attacker has a non-monotonic objective function, we may want to spend more on medium-sized cities than in the monotonic case even at low cost-effectiveness.

6.4 Multi-attribute Terrorist Utility

In this section, we allow factors besides fatalities to affect how "attractive" a target is to terrorists. For example, in addition to evoking terror, terrorists may be interested in generating additional publicity, exacting revenge, achieving specific concessions, causing disorder, or provoking repression (Richardson 2007). The propaganda value of a target may also be considered: "Terrorists seek to attract attention to their cause by employing, or threatening, dramatic acts of violence that capture the attention of the media and terrorize large populations" (Dershowitz 2002). Furthermore, terrorists may be interested in the symbolic value of particular targets; for example, Bin Laden described the Twin Towers as icons of USA "military and economic power" (Rubin and Rubin 2002).

Moreover, Woo (2002) recognized that terrorists consider not only psychological impact (e.g., evoking fear) but also execution difficulty (planning time, required personnel, technical difficulty, and consumption of financial and material resources) when choosing an attack strategy. Similarly, Rosoff and John (2009) found that terrorists consider not only numbers of fatalities and injuries, terror, economic impact, and the symbolic value of their attacks but also the time required to plan an attack, human resources required, and the cost of the attacks. Beitel et al. (2004) likewise consider multiple attacker attributes, including both the resources required for attacks on particular targets, as well as the "return" on those investments, for attacks against targets such as US buildings, US corporate interests, and transportation assets.

Here, we adopt the balanced-scorecard model of Beitel et al. (2004). The "investment" measures considered by Beitel et al. (2004) are human resources (A_1); terrorist resources, such as funding, weapons, explosives, and knowledge (A_2); and terrorist schedule (A_3), or the time required for planning, deployment, and implementation of an attack. The return or "damage" measures include loss of life (A_4), direct economic loss (A_5); national economic stress and inconvenience (A_6), reflecting "impacts on Western lifestyles"; decreased Western presence (A_7); increased "Islamic presence" (A_8); and the opportunity to leverage with other terrorists (A_9). Another measure included in a spreadsheet provided by coauthor Plum (personal communication 2007) accounts for the symbolic values of particular targets (A_{10}).

Beitel et al. (2004) also consider the likelihood of attack success, but we omit this as an attribute, since in the model of Bier et al. (2008), the likelihood of attack success, $p(c_i)$, is dependent on the defensive resources allocated to a given target (c_i)

Table 6.2 Mean values for the normalized exponents w_j

Attribute	Exponent in balanced-scorecard model	Mean normalized exponent for A	Mean normalized exponent for B	Mean normalized exponent for C
w_1	2	0.381		0.073
w_2	2	0.381		0.073
w_3	1.25	0.238		0.046
w_4	2		0.091	0.073
w_5	3		0.136	0.11
w_6	3		0.136	0.11
w_7	6		0.273	0.22
w_8	6		0.273	0.22
w_9	1		0.045	0.037
w_{10}	1		0.045	0.037
Sum		1	1	1

and the cost-effectiveness of that investment (λ), and is not a fixed constant as assumed by Beitel et al.

Beitel et al. used a multiplicative model of target attractiveness, where the exponent w_j of attribute A_j in the balanced-scorecard model reflects the importance of that attribute. Thus, the overall target ratings are of the form

$$A = \text{investment score} = \prod_{j=1}^{3} A_j^{w_j},$$

$$B = \text{return score} = \prod_{j=4}^{10} A_j^{w_j},$$

$$C = \text{total score} = AB = \prod_{j=1}^{10} A_j^{w_j}.$$

In our analysis, we normalize the exponents assumed by Beitel et al. as shown in Table 6.2. (Note that the exponents w_j are not strictly speaking weights in the multiplicative model of target attractiveness, but are equivalent to weights in a corresponding additive model for the logarithm of the attractiveness score).

To highlight the fact that the defender is uncertain about attacker preferences, the values of the exponents are assumed to be random. In particular, we assume that the attribute weights w_j for score C follow a Dirichlet distribution, as given by

$$(w_1, \ldots, w_{10}) \sim \text{Dirichlet}(\alpha_1, \ldots, \alpha_{10}),$$

$$f(w_1, \ldots, w_{10}) = \Gamma(\alpha_0) \prod_{j=1}^{10} \frac{(w_j)^{\alpha_j - 1}}{\Gamma(\alpha_j)},$$

Table 6.3 Numerical values of the attributes A_j for different terrorist targets

Asset	A_1	A_2	A_3	A_4	A_5	A_6	A_7	A_8	A_9	A_{10}
US embassies in foreign lands	2.0	0.33	0.50	0.33	0.58	0.50	1.0	0.10	1.1	1.0
US cultural entities in foreign lands	1.0	0.50	0.50	0.42	0.50	0.50	0.90	0.90	1.5	1.0
US corporate in foreign lands	0.67	0.50	0.50	0.33	0.58	0.70	0.90	0.70	1.2	1.2
US corporate in USA	0.67	0.50	0.50	0.33	0.58	0.70	0.90	0.70	1.2	1.2
Foreign corporate in USA	0.67	0.50	0.50	0.50	0.50	0.30	0.90	0.90	1.5	1.0
Military bases in foreign lands	1.0	0.17	0.25	0.33	0.67	0.70	0.90	0.90	1.0	1.0
Military bases in USA	1.0	0.17	0.25	0.33	0.67	0.70	0.80	0.80	1.0	1.0
US roads and bridges	0.67	0.33	0.50	0.50	0.58	0.70	0.30	0.30	1.0	1.0
US air	0.67	0.50	0.20	0.50	0.67	0.70	0.90	0.30	1.0	1.0
US shipping	0.67	0.50	0.33	0.33	0.67	0.70	0.30	0.30	1.0	1.0

Table 6.4 Asset scores based on normalized exponents (rankings are given in parentheses)

Asset	Investment A	Return B	Total C
US embassies in foreign lands	0.7238 (1)	0.4092 (10)	0.4573 (8)
US cultural entities in foreign lands	0.6511 (2)	0.7358 (2)	0.7192 (1)
US corporate in foreign lands	0.5590 (3)	0.7166 (4)	0.6793 (2)
US corporate in USA	0.5590 (3)	0.7166 (4)	0.6793 (2)
Foreign corporate in USA	0.5590 (3)	0.6974 (6)	0.6687 (4)
Military bases in foreign lands	0.3660 (9)	0.7700 (1)	0.6678 (5)
Military bases in USA	0.3660 (9)	0.7220 (3)	0.6341 (6)
US roads and bridges	0.4771 (7)	0.4304 (8)	0.4397 (9)
US air	0.4495 (8)	0.5924 (7)	0.5623 (7)
US shipping	0.5063 (6)	0.4226 (9)	0.4383 (10)

where $\Gamma(z) = \int_0^\infty t^{z-1}e^{-t}dt, \alpha_j>0 \text{ for } j = 1, \ldots, 10, \alpha_0 = \sum_{j=1}^{10} \alpha_j, w_j \geq 0 \text{ for } j = 1, \ldots,$ 10 and $w_{10} = 1 - \sum_{j=1}^{9} w_j$. (The calculations for the investment score A and return score B are similar, but use only subsets of the attacker attributes.) The choice of the Dirichlet distribution ensures that the attribute weights will sum to one, and makes it possible to vary the extent of defender uncertainty about the attacker attribute weights by changing a single parameter, while leaving the mean attribute weights unchanged. In particular, larger values of the parameter α_0 correspond to lower levels of defender uncertainty.

We now apply the model of Bier et al. (2007) to the attribute values and weights provided by Beitel et al. (2004). Table 6.3 shows the numerical values of the attributes $A_1, \ldots, A_{10}$ for ten different types of assets, including US buildings, US corporate interests, and various types of transportation assets. Finally, Table 6.4 shows the values for the scores A, B, and C, based on the attribute values in Table 6.3 and the mean normalized attribute weights in Table 6.2.

Figure 6.3 provides optimal resource allocations for varying levels of cost-effectiveness of the defender's investments, both for low defender uncertainty

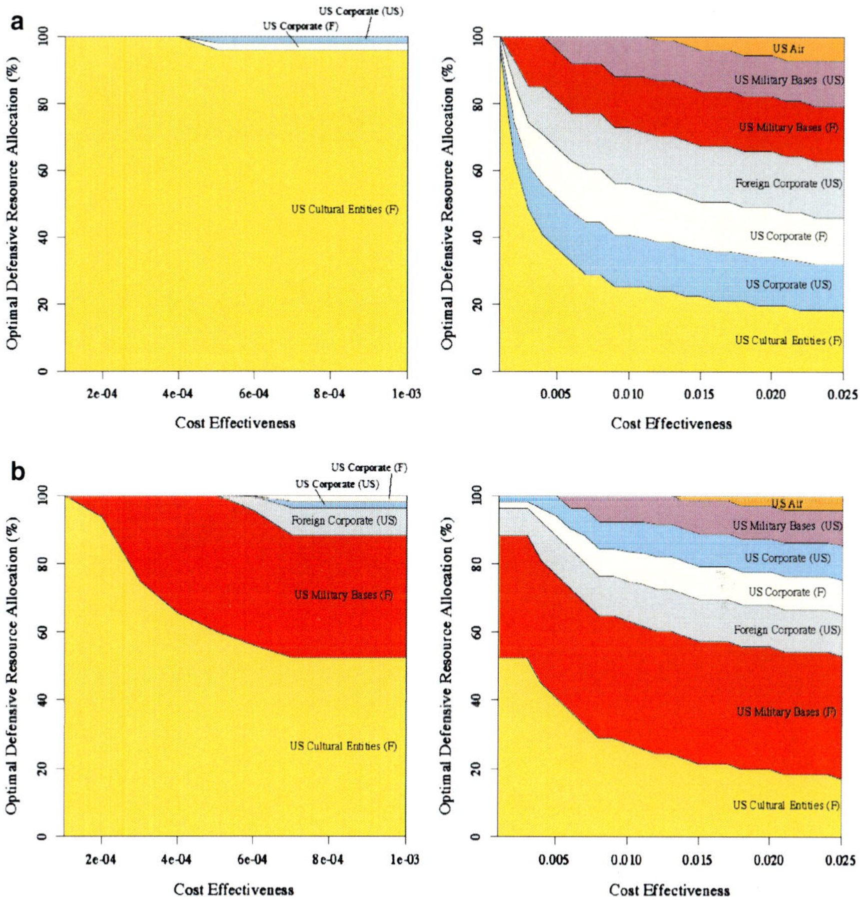

Fig. 6.3 Optimal defensive resource allocations (attacker values targets according to both investment and return measures; defender values targets according to return measures only). (**a**) Low defender uncertainty about attacker preferences; $\alpha_0 = 100$. (**b**) High defender uncertainty about attacker preferences; $\alpha_0 = 10$

($\alpha_0 = 100$; Fig. 6.3a) and for high defender uncertainty ($\alpha_0 = 10$; Fig. 6.3b). In this figure, we assume that the attacker values targets according to C (taking into account both investment and return measures), while the defender values targets according to B (taking into account only return measures).

Figure 6.3 shows more hedging (i.e., more funding of mid-level targets) at high uncertainty (Fig. 6.3b) than at lower levels of uncertainty (Fig. 6.3a) when defensive investments are not highly cost-effective. By contrast, at high levels of cost-effectiveness, we do not observe significantly more hedging with high defender uncertainty. However, the defensive allocations more closely resemble the defender's target valuations at high levels of defender uncertainty. For example,

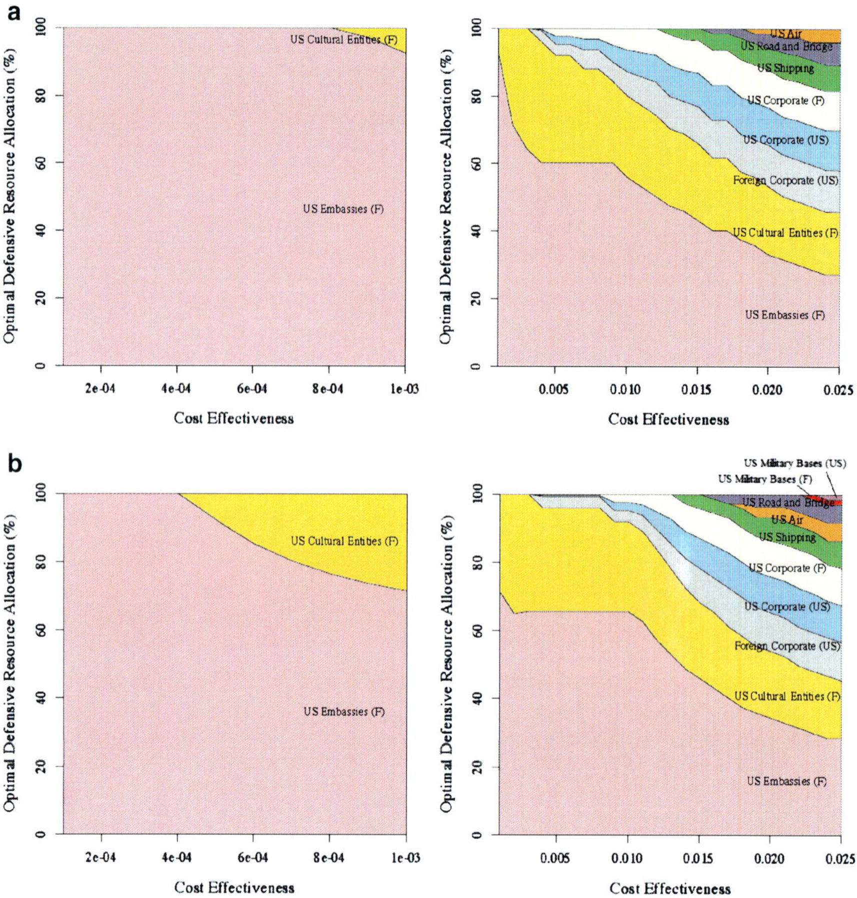

Fig. 6.4 Optimal defensive resource allocations (attacker values targets according to investment measures only; defender values targets according to return measures only). (**a**) Low defender uncertainty about attacker preferences; $\alpha_0 = 100$. (**b**) High defender uncertainty about attacker preferences; $\alpha_0 = 10$

US military bases in foreign lands (ranked first by the defender) get more protection at high levels of defender uncertainty (Fig. 6.3b) than at low level of defender uncertainty (Fig. 6.3a).

However, this model still sometimes yields little or no defensive hedging. For example, consider the case where the attacker is assumed to care about only investment measures (instead of both investment and return measures, as in the previous case), but the defender still cares about only return measures (Fig. 6.4). In this case, two targets (US embassies in foreign lands and US cultural entities in foreign lands) dominate all others in terms of the attacker attributes; see the values for human resources A_1, terrorist resources A_2, and terrorist schedule A_3 in Table 6.3. Therefore, when the cost-effectiveness of defensive investment is low, we see little or no hedging. When the cost-effectiveness of defensive investments is higher, more targets are protected, but there is still no investment in either military bases in foreign lands or military bases in the USA (which are strictly dominated along most attacker attributes), even though both are highly valuable to the defender (ranked first and third by the defender, respectively).

The lack of substantial hedging in this case is not necessarily a problem, since it may not be reasonable to expect high levels of hedging when some targets are strictly dominated on most or all attacker attributes. However, if greater hedging is desired in practice, it may be necessary to find other ways of representing the defender's uncertainty about the attacker's preferences. For example, Wang and Bier (2011) explicitly account for the possibility of unobserved attributes that are important to the attacker but not known to the defender. In this approach, if the defender believes that the attacker puts significant weight on the unobserved attributes, the optimal defensive resource allocation is influenced more heavily by the defender's target valuations, and less heavily by the defender's estimates of the attacker's preferences. Achieving realistic levels of hedging may also require considering uncertainty about more than just the attribute weights—as is done, for example, in the random-utility model by Rosoff and John (2009), where the attribute values A_j are also allowed to be uncertain.

Still another way of incorporating uncertainty into multi-attribute attacker objective functions may be to assess probability distributions directly over the target utilities. However, when the attractiveness of particular targets is inherently correlated, it may be extremely difficult to assess a suitable joint probability distribution over their utilities. Incorporating uncertainty in the attribute weights would thus seem to capture such correlations more naturally.

6.5 Conclusions and Directions for Future Research

Past funding allocations have been criticized as being driven too much by "pork-barrel politics" and too little by risk. The results of simple game-theoretic models with little or no consideration of uncertainty tend to support this view, since these models generally yield protection of only the top few targets, and therefore result in low levels of hedging in defensive investments (i.e., little or no protection of mid-level targets). However, in the real world, decision makers are likely to insist on at least some degree of hedging even at low levels of cost-effectiveness, in order to reflect the fact that they do not know attacker goals perfectly.

Therefore, the main purpose of this chapter was to extend existing game-theoretic models for determining optimal defensive resource allocations in the face of uncertain terrorist preferences and to achieve more realistic levels of defensive hedging. Interestingly, the results of our model suggest that levels of hedging similar to those observed in DHS' actual funding allocations may not be the result of pork-barrel politics alone but may also reflect prudent risk management, especially if there is significant uncertainty about terrorist preferences.

We considered two extensions to previous game-theoretic models: non-monotonic attacker objective functions and multi-attribute attacker objective functions. Non-monotonic attacker objective functions do typically yield greater hedging, in the form of greater protection of mid-level cities even at low levels of cost-effectiveness. Multi-attribute models with uncertain attribute weights also yield hedging in some cases, but this is not always true, especially when some targets are strictly dominated along most or all attacker attributes. This suggests that allowing for uncertain attribute weights in multi-attribute attacker objective functions may not always be sufficient to yield significant hedging at optimality.

Of course, better methods for quantifying key parameters in these models are still needed before the models are ready for application. For example, formal methods of expert elicitation, especially ranking-based methods such as probabilistic inversion (Cooke and Misiewicz 2007; Neslo et al. 2008) or SMARTER (Edwards and Barron 1994), could lead to more justifiable parameter estimates, while overcoming the reluctance of some intelligence experts to express their knowledge quantitatively.

Additionally, we currently lack good metrics for the cost-effectiveness of defensive investments. In recent research, Jamshidi and Bier (2009) attempt to estimate the cost-effectiveness of security expenditures using regression analysis and examine how the reduction in estimated risk depends on the extent of defensive investment. However, further research to better quantify the cost-effectiveness of defensive investments would be extremely useful, especially since cost-effectiveness clearly has a large effect on optimal resource allocations.

References

Asaba N (2006) Concepts and challenges in using risk management in a homeland security setting. Presented at the Pacific Northwest Intergovernmental Audit Forum. http://www.auditforum.org/speaker%20presentations/pacific/pniaf%2009%2006/asaba.pdf. Accessed 21 Apr 2008

Beitel G, Gertman D, Plum M (2004) Balanced scorecard method for predicting the probability of a terrorist attack. Idaho National Engineering Environmental Laboratory, Idaho Falls

Bier VM, Oliveros S, Samuelson L (2007) Choosing what to protect strategic defensive allocation against an unknown attacker. J Public Econ Theory 9(4):563–587

Bier VM, Haphuriwat N, Menoyo J, Zimmerman R, Culpen A (2008) Optimal resource allocation for defense of targets based on differing measures of attractiveness. Risk Anal 28(3):763–770

Brunet A (2006) Vulnerabilities to terrorism. Paper presented at the annual meeting of The Law and Society Association. http://www.aeaweb.org/annual_mtg_papers/2007/0106_0800_1501.pdf. Accessed 22 Apr 2008

Cooke RM, Misiewicz J (2007) Discrete choice with probabilistic inversion: application to energy policy choice and wiring failure. Presented at Mathematical Methods in Reliability, July

Davis R (1998) Combating terrorism: observations on the Nunn-Lugar-Domenici Domestic Preparedness Program. Government Accountability Office Testimony before the Subcommittee on National Security, International Affairs and Criminal Justice, Committee on Government Reform and Oversight, House of Representatives

Dershowitz A (2002) Why terrorism works: Understanding the threat, responding to the challenge. Yale University Press

Edwards E, Barron FH (1994) SMARTS and SMARTER: improved simple methods for multiattribute utility measurement. Organ Behav Hum Decis Process 60(3):306–325

Government Accountability Office (2008) Homeland security grant program risk-based distribution methods. Presentation to Congressional Committees–November 14, and December 15, GAO-09-168R, Dec 23

Guikema SD (2009) Game theory models of intelligent actors in reliability analysis: an overview of the state of the art. In: Bier VM, Azaiez MN (eds) Game Theoretic Risk Analysis of Security Systems, Springer Science+Business, New York

Jamshidi T, Bier VM (2009) Cost effectiveness of investments in defense of critical infrastructure. Society for Risk Analysis Annual Meeting, Baltimore, Maryland, December 6–9

Masse T, O'Neil S, Rollins J (2007) The Department of Homeland Security's risk assessment methodology: evolution, issues, and options for congress. CRS Report for Congress

National Research Council (2008) Department of Homeland Security bioterrorism risk assessment: a call for change. http://books.nap.edu/openbook.php?record_id=12206. Accessed 1 Oct 2010

Neslo R, Micheli F, Kappel CV, Selkoe KA, Halpern BS, Cooke RM (2008) Modeling stakeholder preferences with probabilistic inversion: application to prioritizing marine ecosystem vulnerabilities. In: Linkov I, Ferguson E, Magar V (eds) Real time and deliberative decision making: application to risk assessment for non-chemical stressors. Springer, Amsterdam, pp 265–284

Parnell G, Borio L, Brown G, Banks D, Wilson A (2008) Scientists urge DHS to improve bioterrorism risk assessment. Biosecur Bioterror 6(4):353–356

Reese S (2006) Homeland security grants: evolution of program guidance and grant allocation methods. CRS Report for Congress

Richardson L (2007) What terrorists want: understanding the enemy, containing the threat. Random House Trade Paperbacks, Reprint Edition, New York

Ripley A (2004) How safe are we? How we got homeland security wrong. Time, March 29

Rosoff H, John R (2009) Decision analysis by proxy for the rational terrorist. Workshop on Security and Human Behaviour, Boston, 11–12 June 2009

Rubin B, Rubin J (eds) (2002) Anti-American terrorism and the Middle East. Oxford University Press, New York

Sandler T, Siqueira K (2009) Games and terrorism: recent developments. Simul Gaming 40 (2):164–192

Walker D (2002) Homeland security: responsibility and accountability for achieving national goals. Government Accountability Office Testimony before the Committee on Governmental Affairs, U.S. Senate

Wang C, Bier VM (2011) Target-hardening decisions based on uncertain multi-attribute terrorist utility. Decision Anal 8:286–302

Willis H (2005) Analyzing terrorism risk. Written testimony submitted to the House Financial Services Committee, Subcommittee on Oversight and Investigations, and the House Homeland Security Committee, Subcommittee on Intelligence, Information Sharing, and Terrorism Risk Assessment. Text from: RAND Cooperation Testimony Series

Woo G (2002) Quantifying insurance terrorism risk. Prepared for the National Bureau of Economic Research meeting. Cambridge, Massachusetts. http://www.rit.edu/cos/math/cmmc/literature/Woo_2002b.pdf. Accessed 2 Oct 2012

Chapter 7
Mitigating the Risk of an Anthrax Attack with Medical Countermeasures

Jeffrey W. Herrmann

Abstract This chapter presents a simulation model that can be used to prepare for a bioterrorism attack that releases anthrax spores and exposes thousands of persons to this deadly disease. The model predicts the expected number of deaths using information about the size of the population, the number exposed, the progress of the disease, the resources available for distributing medication and treating the ill, and the size of local medication stockpiles. The chapter also presents a risk management approach for allocating a limited medication stockpile to multiple cities to minimize the expected number of deaths. The results show that the optimal allocation can be quite different from allocations that are proportional to population size.

7.1 Introduction

This chapter presents the combined use of two different operations research (OR) models to address a risk management problem in the context of preparing for a bioterrorism attack. The first model is a predictive model that estimates the number of deaths based on information about the attack scenario. The second model is an optimization model that can determine the optimal allocation of limited resources.

The deliberate release of aerosolized anthrax spores in a large city will expose many thousands of residents to this deadly disease. Promptly distributing medical countermeasures (antibiotics) to those exposed is a key step in preventing illness and deaths. Avoiding delays in this distribution is critical, but such a response will require enormous resources. State and local health departments have developed contingency plans for points of dispensing (PODs), the primary distribution channel, and other countermeasures and strategies have been proposed and tested, such as

J.W. Herrmann (✉)
A. James Clark School of Engineering, University of Maryland, College Park, MD 20742, USA
e-mail: jwh2@umd.edu

J.W. Herrmann (ed.), *Handbook of Operations Research for Homeland Security*, International Series in Operations Research & Management Science 183,
DOI 10.1007/978-1-4614-5278-2_7,

employing the U.S. Postal Service to deliver antibiotics directly to residences (cf. Executive Order 13527), and prepositioning (CDC 2011). Prepositioned medications include forward-deployed (local) stockpiles, workplace and hospital caches, and predispensed medical countermeasures that are stored by heads of households (IOM 2012).

In this chapter, the term "medical countermeasure" (MCM) is used to refer to a regimen of antibiotics that will be stored in a local stockpile, a hospital or workplace cache, or predispensed. After an attack, MCMs in a local stockpile or a hospital or workplace cache will be distributed in PODs to those who believe that they were exposed. A community's prepositioning strategy includes its local stockpile and predispensed MCMs.

In 2006 the Centers for Disease Control and Prevention (CDC) and the Missouri Department of Health and Senior Services distributed prototype MedKits (a proposed type of predispensed MCM) to over 4,000 households, and 97% of the households returned their kits after 2–8 months. A majority of the participants stated that they would like to have a MedKit to keep in their home, and most would be willing to pay for the MedKit (CDC 2007).

Although Richter and Khan (2009) and Zaric et al. (2008) described studies comparing some of the various strategies, these studies did not consider the impact of predispensing MCMs, and we are unaware of any available models that can show how predispensing MCMs would mitigate the consequences of an anthrax attack.

Models that can predict this benefit for particular scenarios should be valuable to public health officials who are considering whether and to what extent to predispense MCMs.

Of course, the usefulness of any model depends upon the availability of good data. Because data can provide only estimates of important values, it is important to conduct some sensitivity analysis to determine how the results could vary.

This chapter considers this issue. First, it discusses a model that estimates the expected number of deaths that result from an anthrax attack in a community that has adopted a specific prepositioning strategy (both predispensed MCMs and local stockpiles to be distributed at PODs). This model could be used by public health officials in a particular urban area (city) for understanding how prepositioning MCMs could reduce the expected number of deaths in that city. Then this chapter presents an approach that uses this model in order to determine the optimal allocation of a limited store of MCMs to multiple cities. In other words, first we consider the problem of defending one city, and then we consider the problem of defending multiple cities.

Predispensing MCMs has been advocated because it could reduce the number of persons who would go to PODs and reduce the time needed to distribute medication (Bicknell 2003; IOM 2008). Moreover, individuals are interested in preparing for emergencies and being able to take care of themselves in an emergency (National Biodefense Science Board 2008). Nevertheless, there are concerns about the inappropriate use of MCMs, which might further the antibiotic resistance of more common bacterial infections, the risks of overdoses, adverse effects, and using expired drugs, the liability of any agency that promotes off-label use of medication, and the challenge of satisfying Food and Drug Administration (FDA) regulations (National Biodefense Science Board 2008; Troy 2010).

The decision to preposition MCMs must therefore consider a variety of issues. Those who must pay for the MCMs will consider their cost. Others will consider the cost of treating those who become ill, the cost of maintaining local stockpiles, and other costs (Zaric et al. 2008). Health officials will consider the risks of misuse. In general, if many thousands of people begin taking antibiotics (from predispensed MCMs and from PODs), there will be adverse impacts (Shepard et al. 2002). Ideally, analysis like that described here would be combined with deliberation about the legal, regulatory, safety, ethical, and cost considerations (which are beyond the scope of this chapter) in a risk-informed decision-making process (Stern and Fineberg 1996). This chapter focuses on the problem of predicting how predispensing MCMs will reduce the expected number of deaths.

The IOM emphasized that jurisdictions should evaluate which prepositioning strategies are appropriate for the community, proposed a decision-making framework, and recommended additional research in this area (IOM 2012). Centralized stockpiles (like the Strategic National Stockpile) cost less, have more flexibility, and eliminate the potential for misuse; but forward-deployed (local) stockpiles, workplace and hospital caches, and predispensed MCMs (personal stockpiles and MedKits) reduce the time needed to distribute prophylaxis, which could reduce the number of deaths in an anthrax attack. There are also legal, regulatory, and ethical aspects to consider.

Within this framework, estimating the health benefits (e.g., the reduction in the expected number of deaths from an attack) is a key attribute, but this is difficult to assess (IOM 2012); this chapter presents a model for estimating this attribute. Although exercises are important tools for training and assessment (and valuable sources of data), full-scale exercises are expensive and disruptive, so models that can predict the impact of an MCM strategy are essential for evaluating prepositioning strategies.

This study considers only predispensing MCMs to the general population. It may be desirable to predispense MCMs to first responders and other personnel who are essential to continuity of operations. In general, the allocation of scarce resources to groups within an urban area must be considered within a framework of ethical guidelines that emphasize the relevant moral principles (Kinlaw et al. 2009).

The approach described in this chapter uses a compartmental model that predicts the deaths and hospitalizations from an anthrax attack. The model is based on a model described by Zaric et al. (2008), which focused on PODs and did not consider the impact of predispensing MCMs. The disease progression model used here is based upon a review of all of the cases of inhalational anthrax that have been published since 1900 (Holty et al. 2006).

The objective of the work described in this chapter was to develop a model that could estimate the impact of predispensing MCMs in a community and to use that model to consider some possible scenarios to get some insight into the effectiveness of predispensing MCMs. This chapter will briefly describe the model and the results. A complete description of the mathematical model was provided by Houck and Herrmann (2011). After discussing this model, the chapter will present an approach that uses the model to determine the optimal allocation of a limited store of MCMs to multiple cities.

7.2 MCM Modeling Approach

For this study, we developed a compartmental model that includes both the progression of the disease and the logistics of treatment. A compartmental model represents the flows of individuals between compartments. Each compartment represents a number of homogeneous individuals (that is, they are identical with respect to their condition and treatment status). The compartments are mutually exclusive and collectively exhaustive. Compartmental models can be analyzed using differential equations (which can be solved exactly in some situations) or difference equations (which we will use here). Compartmental models have been used in studies of biology, medicine, ecosystems, populations, and economics (Jacquez 1996; Walter and Contreras 1999).

The model used here (the "MCM model") extends the model presented by Zaric et al. (2008) (the "POD model") by adding additional compartments for the population that has predispensed MCMs and revising the transition equations used to predict the flow between compartments. These revisions include improvements to the published equations and a scheme for prioritizing the transitions.

The MCM model has 28 compartments to represent distinct groups within a population. The primary distinctions between the compartments are the exposure of the individuals, the progression of the disease (including the definitive conditions of death or recovery), the treatment status, and the possession of MCMs. The model includes the 21 compartments of the POD model and adds seven more: one for death, two for potential exposures who possess predispensed MCMs, three for exposed individuals who possess predispensed MCMs, and one for those who adhere to prophylaxis (and cannot become ill). Table 7.1 lists all 28 compartments. Figure 7.1 is a schematic that shows the possible flows in the MCM model.

The model includes two types of flows. The first type of flow corresponds to changes in the disease in those who were exposed. The transition path is from the incubation stage to the prodromal stage, to the fulminant stage, and then to death. The second type of flow corresponds to changes in awareness and treatment. Exposed persons and potential exposures are first unaware of their exposure or the need for prophylaxis. When they become aware, they seek prophylaxis (essentially, they are in a queue for prophylaxis). Exposed persons who become sick then seek treatment and receive treatment. Those who are treated may recover.

Because many of the compartments have multiple outflows (corresponding to different transitions), it is important to define the relative priority of the transitions. In the MCM model, the highest priority transitions are those that correspond to the progression of the disease, recovery, and death. The second priority transitions are those that correspond to awareness, prophylaxis, and treatment. The number who

Table 7.1 The compartments in the MCM model

Compartment	Exposure, treatment status	Progression of disease	Treatment began	MCMs?
1	Exposed, unaware of exposure	Incubation	n.a.	No
2	Exposed, unaware of exposure	Prodromal	n.a.	No
3	Exposed, unaware of exposure	Fulminant	n.a.	No
4	Aware of exposure	Incubation	n.a.	No
5	Aware of exposure	Prodromal	n.a.	No
6	Aware of exposure	Fulminant	n.a.	No
7	Not adhering	Incubation	Prophylaxis begun in incubation stage	Both
8	Not adhering	Prodromal	Prophylaxis begun in incubation stage	Both
9	In prophylaxis	Prodromal	Prophylaxis begun in prodromal stage	Both
10	Not adhering	Fulminant	Prophylaxis begun in incubation stage	Both
11	In prophylaxis	Fulminant	Prophylaxis begun in prodromal stage	Both
12	In prophylaxis	Fulminant	Prophylaxis begun in fulminant stage	Both
13	In treatment	Prodromal	Treatment begun in prodromal stage	Both
14	In treatment	Fulminant	Prophylaxis or treatment begun in prodromal stage	Both
15	In treatment	Fulminant	Prophylaxis begun in incubation stage, treatment begun in fulminant stage	Both
16	In treatment	Fulminant	No prophylaxis or prophylaxis begun in fulminant stage	Both
17	Exposed	Recovered	n.a.	Both
18	Not exposed	n.a.	n.a.	Both
19	Potential exposure, not seeking prophylaxis	n.a.	n.a.	No
20	Potential exposure, seeking prophylaxis	n.a.	n.a.	No
21	Potential exposure, in prophylaxis	n.a.	n.a.	No
22	Exposed	Dead	n.a.	Both
23	Potential exposure, not seeking prophylaxis	n.a.	n.a.	Yes
24	Potential exposure, in prophylaxis	n.a.	n.a.	Yes
25	Exposed, unaware of exposure	Incubation	n.a.	Yes

(continued)

Table 7.1 (continued)

Compartment	Exposure, treatment status	Progression of disease	Treatment began	MCMs?
26	Exposed, unaware of exposure	Prodromal	n.a.	Yes
27	Exposed, unaware of exposure	Fulminant	n.a.	Yes
28	Adhering to prophylaxis	Prophylaxed	Prophylaxis begun in incubation stage	Both

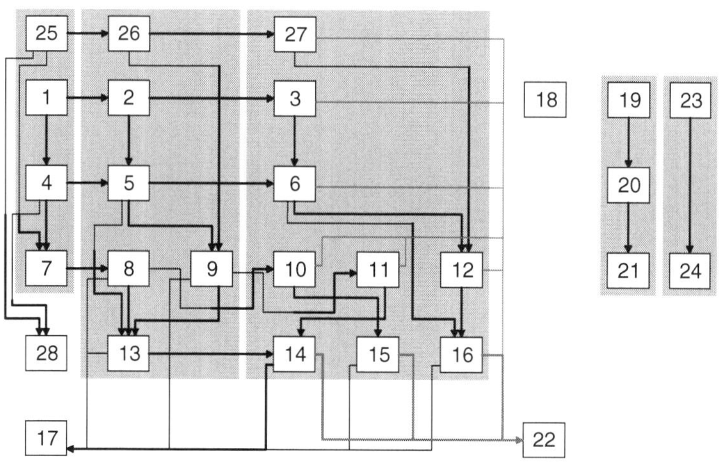

Fig. 7.1 Schematic of the flows between compartments in the MCM model

can be given prophylaxis is limited by the prophylaxis dispensing capacity, which is spread proportionally across all of the compartments with persons waiting for prophylaxis. Likewise, the number who can be treated is limited by treatment capacity, which is spread proportionally across all of the compartments with persons waiting for treatment.

Like the POD model, the MCM model is a deterministic, discrete-time model. The time period is 1 h. Let $X_i(t)$ be the expected number of individuals in compartment i at time t. The model calculates these values for a 2400-hour (100-day) time horizon. At each point in time, we calculate the transitions, where $\phi_{ij}(t)$ is the expected number of individuals who move from compartment i to compartment j due to illness, recovery, or death, and $\Psi_{ij}(t)$ is the expected number of individuals who move from compartment i to compartment j due to awareness, prophylaxis, or treatment. $Y_i(t)$ is the expected number of individuals in compartment i at time t after the transitions due to illness, recovery, and death, and these values are used to determine the values of $\Psi_{ij}(t)$, as described in detail by Houck and Herrmann (2011). Note that many of the transitions are always zero; the notation used here allows a compact expression of the transition equations, which can be written as follows:

$$Y_j(t) = X_j(t) + \sum_{i=1}^{28} \phi_{ij}(t) - \sum_{i=1}^{28} \phi_{ji}(t),$$
$$X_j(t+1) = Y_j(t) + \sum_{i=1}^{28} \Psi_{ij}(t) - \sum_{i=1}^{28} \Psi_{ji}(t). \tag{7.1}$$

In general, during each time period, a certain fraction of the individuals in a compartment will move to another compartment. This flow is governed by the values of parameters describing the disease progression (the rates at which individuals become ill, recover, or die) and the prophylaxis and treatment capacity. For a given flow, the fraction may be time-invariant (like a death rate) or may change over time as available treatment capacity changes. Many of the parameters given below and used in the MCM model are the same as those used in the POD model, but the model can be changed easily to consider scenarios with other parameters. A spreadsheet version of the model has been constructed and is available from the author.

7.2.1 Attack and Response Timeline

The consequences of an anthrax attack depend upon the attack scenario, including the amount and characteristics of the anthrax spores that are released, the time and location of the attack, the weather, the number and condition of people who come into contact with the spores, and the speed of detecting, investigating, and responding to the attack (Graham et al. 2008; Oren 2009). The MCM model can be used for a wide range of scenarios, but those considered in this study were based on the following timeline. The attack occurs at $t = 0$ h. The attack is detected at $t = 48$ h. Local stockpiles of both intravenous antibiotics (IVs) for treatment and antibiotics for dispensing will become available 5 h later at $t = 53$ h. Intravenous antibiotics (IVs) for treatment, antibiotics for dispensing, and additional ventilators from the push pack will become available 16 or 28 h after attack detection at $t = 64$ or 76 h. (This is due to a 12- or 24-h delay in receiving the push pack and another 4-h delay in getting the material from the push pack ready.) Intravenous antibiotics (IVs) for treatment and antibiotics for dispensing from vendor-managed inventory (VMI) will become available 36 h after attack detection at $t = 84$ h. At $t = 96$ h (48 h after attack detection), complete POD capacity will be available. In the MCM model, these times are all parameters that can be changed to consider other scenarios.

7.2.2 Exposure

The population consists of three large subpopulations: those who were exposed to the anthrax attack, those who were not exposed to the anthrax attack (or inhaled too few anthrax spores to become ill), and those who believe that they may have been exposed (because of their proximity to the attack or for other reasons). The persons in this last group, called "potential exposures," will undergo prophylaxis (by going to PODs and taking their MCMs) but cannot become ill. Like the POD model, the MCM model does not divide the population by age because we assume that the progression of anthrax does not depend upon age. We assume that exposure to the attack is independent of MCM possession. Therefore, the proportion who possesses predispensed MCMs is the same in all three subpopulations. Because those who were not exposed cannot become ill and do not seek prophylaxis whether they have MCMs or not, the model treats all of these persons in one compartment. Those who are exposed are initially unaware of their exposure. Some become aware during incubation, some when they are prodromal, and some when they are fulminant.

7.2.3 Disease Progression

Inhalational anthrax begins after aerosolized anthrax spores pass into the lungs, germinate, and begin replication (Oren 2009). The disease progresses from the incubation stage to the prodromal stage, to the fulminant stage, and then to death. An exposed person in the incubation stage is infected with anthrax but is asymptomatic. A person in the prodromal stage has nonheadache neurological symptoms (e.g., fever, muscle aches, and fatigue) that are similar to flu. A person in the fulminant stage is severely ill, has respiratory distress, and may die within days. We assume that only persons in the fulminant stage can die. Because it considers a short period of time, this model does not consider any other causes of death. Those in the prodromal and fulminant stages who start treatment may recover. The rates at which persons become ill, recover, or die vary based on their status. The times to become ill, recover, or die are modeled as geometric distributions (thus, they are memoryless), and the probability of this event per time unit is the reciprocal of the mean time. Table 7.2 gives these probabilities, which are based on the expected times in Zaric et al. (2008) and Holty et al. (2006). For example, when the expected time to become fulminant is 122.4 h, the probability (each hour) of becoming prodromal is 1/122.4. As in the POD model, in the MCM model some who begin prophylaxis may recover from the prodromal stage without treatment in an ICU.

7.2.4 Awareness and Prophylaxis

Those who become ill can become aware at any time: for those who are in the prodromal stage, the probability, each hour, of becoming aware equals 1/72; for

Table 7.2 Fraction of each compartment that becomes ill, recovers, or dies each hour

Compartment	Become prodromal	Become fulminant	Die	Recover
1	$p(t)$			
2		$\theta_2(t)\gamma + (1 - \theta_2(t))\eta$		
3			1/26.4	
4	$p(t)$			
5		$\theta_5(t)\gamma + (1 - \theta_5(t))\eta$		
6			1/26.4	
7	$p(t)$			
8		1/122.4		1/21.7
9		1/122.4		1/21.7
10			1/26.4	
11			1/26.4	
12			1/38.4	
13		1/122.4		1/21.7
14			1/24	1/720
15			1/24	1/720
16			1/38.4	1/720
25	$p(t)$			
26		$\theta_{26}(t)\gamma + (1 - \theta_{26}(t))\eta$		
27			1/26.4	

those who are in the fulminant stage, the probability equals 1/48. After the attack is detected, for those who are in the incubation stage and for potential exposures, the probability, each hour, of becoming aware equals 1/72. This awareness process is a special case of the Bass diffusion process, and the rates are the same as those in the POD model (Zaric et al. 2008).

Because they do not need to go to PODs, those who have MCMs and were exposed (and potential exposures who have MCMs) can start prophylaxis as soon as they become aware. However, some of those who had MCMs may be unable to use them, so we assume that 5% of those given MCMs do not have them at the time of the attack.

Those who do not have MCMs will go to PODs, where the prophylaxis dispensing consists of an oral antibiotic, either ciprofloxacin or doxycycline. Prophylaxis dispensing capacity is limited. It depends upon the facilities and staff available. In the scenarios considered in this chapter, we assume that there is a fixed maximum prophylaxis dispensing capacity, which is 5,833.33 persons per hour. Prophylaxis dispensing is also limited by the availability of medication. Although a complete regimen has 60 days of medication, we assume that only 14-day abbreviated regimens are dispensed until the VMI becomes available. We assume that the local stockpile has 50,000 doses (3,571 abbreviated regimens), the push pack provides 2,718,000 doses (194,143 abbreviated regimens), and the VMI provides sufficient doses for everyone to receive a complete regimen and enough IV antibiotics for everyone who needs them.

Those who adhere to their prophylaxis will not become ill, but some who begin prophylaxis during the incubation stage will not adhere and may become ill. (We assume that those who begin prophylaxis in the prodromal and fulminant stages will always adhere.) We considered two different adherence rates: 65% and 90%. During the 2001 anthrax attack in Washington, D.C., only 64% of those who received prophylaxis adhered fully to prophylaxis (Shepard et al. 2002). Other studies assumed that 90% of those who receive prophylaxis will adhere (Wein et al. 2003). If the regimens in the MCMs and those dispensed at the PODs are not complete regimens, then these persons will need to obtain the remainder. We assume that the process of dispensing the remainder will be done on a less urgent basis and will not interfere with primary dispensing capacity.

7.2.5 Treatment

Persons who become ill need treatment, which consists of three antibiotics administered intravenously in an intensive care unit (ICU). All who begin treatment adhere to it. Treatment capacity is limited by the availability of IV antibiotics, ventilators, respiratory technicians, and ICU beds. We assume that the local stockpile has 500 days of IV antibiotics, the push pack provides 21,492 days of IV antibiotics, and the VMI provides sufficient IV antibiotics for everyone who is being treated. We assume that 100 ventilators are available when the attack occurs, and the push pack provides 100 more. We assume that each respiratory technician can monitor 10 patients, 200 respiratory technicians are available when the attack occurs, and 2,000 ICU beds are available when the attack occurs. There must be at least one day's worth of IV antibiotics to begin treatment of one patient.

7.2.6 Antibiotic Inventory

The model also keeps track of the inventory of antibiotics available for prophylaxis and the IVs available for treatment. Each hour, the inventory of antibiotics (days of medication) is reduced by the number of days of medication dispensed (the product of the number who receive prophylaxis and the number of days in the dispensed regimen at that time). In addition, the inventory of IVs (days of IVs) is reduced by the number of those being treated divided by 24. The inventory for antibiotics and IVs increases when the local inventory becomes available (at $t = 53$ h), when the push pack inventory is available and ready, and when the VMI becomes available (at $t = 84$ h). The number of ventilators also increases when the push pack inventory is available and ready.

7.2.7 *Model Output*

The model determines, for a particular scenario and a given number of predispensed MCMs, the expected number of persons who die from anthrax within 2,400 hours of the attack. From this, we calculate the mortality rate by dividing the expected number of deaths by the number exposed in that scenario.

7.3 MCM Model Results

To illustrate the use of the model to evaluate how predispensing MCMs reduces the risk of an anthrax attack (by reducing the expected number of deaths), we considered a set of 36 scenarios like those considered by Zaric et al. (2008). The population was five million people in all scenarios (this is approximately the population of metropolitan Philadelphia). To create the scenarios, we varied the number of people exposed, the fraction of unexposed people who believe they were exposed (and thus use medication that could be reserved for exposed people), the time until the push pack becomes available, and the prophylaxis adherence rate. We considered attack scenarios with 50,000 exposed, 500,000 exposed, and 1,250,000 exposed. Rates of 0.01, 0.1, and 0.5 were used for the fraction of the unexposed population seeking prophylaxis. We used prophylaxis adherence rates of 65% and 90%. Values of 12 h and 24 h were used for push pack availability. Table 7.3 lists the characteristics of the 36 scenarios.

For each scenario, we varied the number of predispensed MCMs from 0 to 5 million and used the MCM model to estimate the expected number of deaths. Figure 7.2 shows the mortality rates as a function of the number of predispensed MCMs for six scenarios (14, 28, 22, 26, 30, and 34).

Predispensing MCMs significantly reduces the expected number of deaths and the expected mortality rate. A key factor that affected the size of the decrease was the rate at which those who were not exposed believed that they were (which determined the number of potential exposures). When the number of potential exposures was large, then medication was not immediately available for those who truly needed it, which delayed their prophylaxis. In these scenarios, predispensing more MCMs saved many lives. For example, in scenario 36, 1,250,000 people were exposed and another 1,875,000 were potential exposures, the adherence rate was 65%, and the push pack availability was 24 h. In this scenario, predispensing five million MCMs reduced the expected number of deaths from 621,907 to 140,368. When the number of potential exposures was small, the antibiotics were going to those who needed it with minimal delay, and predispensing more MCMs had minimal impact on the number of deaths.

Regardless of the number of predispensed MCMs, it appears that some number of deaths is unavoidable. For example, in scenario 1, which had the smallest attack (50,000 persons exposed), the smallest number of potential exposures (1% of those

Table 7.3 Characteristics of each scenario

Scenario	Number exposed	Number of potential exposures	Adherence rate (%)	Push pack availability (hours)
1	50,000	49,500	90	12
2	50,000	49,500	90	24
3	50,000	49,500	65	12
4	50,000	49,500	65	24
5	50,000	495,000	90	12
6	50,000	495,000	90	24
7	50,000	495,000	65	12
8	50,000	495,000	65	24
9	50,000	2,475,000	90	12
10	50,000	2,475,000	90	24
11	50,000	2,475,000	65	12
12	50,000	2,475,000	65	24
13	500,000	45,000	90	12
14	500,000	45,000	90	24
15	500,000	45,000	65	12
16	500,000	45,000	65	24
17	500,000	450,000	90	12
18	500,000	450,000	90	24
19	500,000	450,000	65	12
20	500,000	450,000	65	24
21	500,000	2,250,000	90	12
22	500,000	2,250,000	90	24
23	500,000	2,250,000	65	12
24	500,000	2,250,000	65	24
25	1,250,000	37,500	90	12
26	1,250,000	37,500	90	24
27	1,250,000	37,500	65	12
28	1,250,000	37,500	65	24
29	1,250,000	375,000	90	12
30	1,250,000	375,000	90	24
31	1,250,000	375,000	65	12
32	1,250,000	375,000	65	24
33	1,250,000	1,875,000	90	12
34	1,250,000	1,875,000	90	24
35	1,250,000	1,875,000	65	12
36	1,250,000	1,875,000	65	24

not exposed, which is 49,500 in this case), a 90% adherence rate, and a 12-h push pack availability, the estimated number of deaths was 4,102 (a mortality rate of 8.2%) even when MCMs were predispensed to the entire population. The unavoidable deaths result from the delays in detecting the attack and starting prophylaxis (during which time some exposed persons become very ill), the loss of MCMs among those who received them, and the imperfect adherence rate.

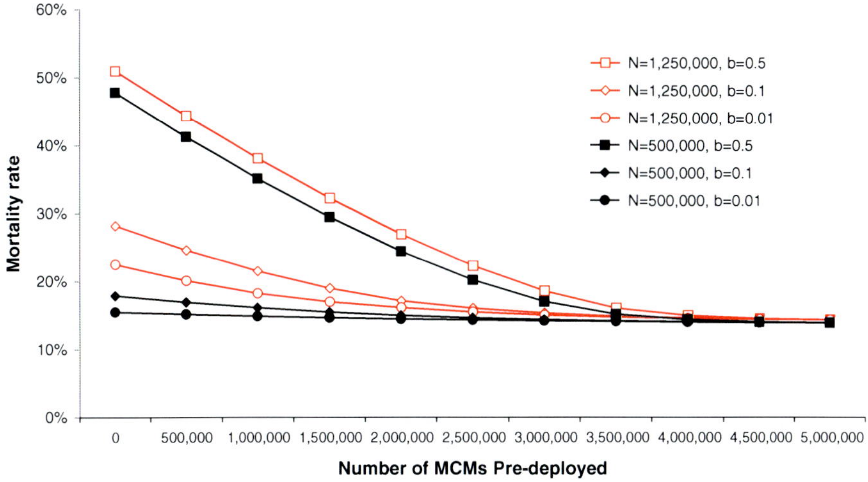

Fig. 7.2 The mortality rate decreases when more MCMs are distributed before an attack. The results shown are for scenarios in which the prophylaxis adherence rate is 90 % and the push pack delay is 24 h. *N* is the number exposed. *b* is the fraction of nonexposed persons who will seek prophylaxis (potential exposures)

Given a quantity of MCMs, a jurisdiction could decide to keep the MCMs in its local stockpile instead of predispensing them. Which would affect mortality rates more? We compared, for the scenarios in Table 7.3, the following possibilities (1) no additional MCMs beyond the local stockpile of 50,000 regimens, (2) predispensing 1,000,000 MCMs, and (3) adding 1,000,000 MCMs to the local stockpile (with no predispensing). We then compared the mortality rates in these three cases. The results showed that, in these scenarios, adding MCMs to the local stockpile did not reduce the expected number of deaths as much as predispensing MCMs. Indeed, the average reduction in the mortality rate when adding MCMs to the local stockpile was 0.004. The average reduction in the mortality rate when predispensing MCMs was less than 0.117. Figure 7.3 shows the mortality rates for nine of the 36 scenarios. As the number of people exposed and the number of potential exposures increase, predispensing MCMs causes a larger reduction in the mortality rate, but adding MCMs to the local stockpile hardly changes the mortality rate.

To estimate the sensitivity of the MCM model to the input parameters, these were grouped into the following sets: awareness, local, times, push pack, disease, and recovery. Instead of adjusting each value individually, all of the values in the group were modified simultaneously and by the same amount.

In particular, the "awareness" group included all of the average times that uninformed persons become aware of the attack. These were first increased by 50% and then reduced by 50%.

The "local" group included all of the local capabilities, including the local stockpiles of antibiotics, the number of technicians, ventilators and ICU beds, and the maximum POD capacity. These were increased by 50% and reduced by 50%.

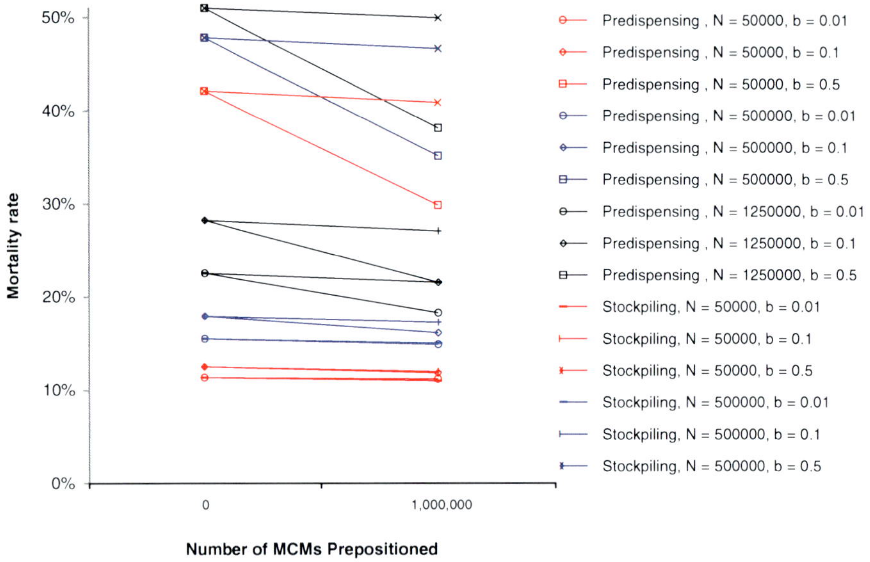

Fig. 7.3 Mortality rate for the scenarios in which the prophylaxis adherence rate is 90 % and the push pack delay is 24 h. For each scenario, three rates are shown: when no MCMs are prepositioned, when 1,000,000 MCMs are predispensed ("predispensing"), and when 1,000,000 MCMs are added to the local stockpile ("stockpiling"). *N* is the number of people exposed, and *b* is the fraction of nonexposed persons who will seek prophylaxis (potential exposures)

The "times" group included all of the times involved in detecting the attack and responding to it. These were increased by 50% and reduced by 50%. Note that the default value for the time until the push pack arrived was set to 24 h.

The "push pack" group included the parameters describing the size of the push pack, including the number of regimens, the number of IV antibiotics, and the number of ventilators. These were increased by 50% and reduced by 50%.

The "disease" group included all of the rates and times associated with the progression of the disease. We first considered the "worst case" by reducing the expected incubation time, the expected time to become fulminant, and the expected time to die by 50%. We then considered the "best case" by doubling the expected incubation time, the expected time to become fulminant, and the expected time to die.

The "recovery" group included all of the average times associated with recovering from the disease. These were increased by 50% and reduced by 50%.

Finally, we changed to 50% (from 95%) the parameter p_M the probability that someone with a regimen of predispensed MCMs would still have the MCMs when told to take them after an attack.

We evaluated a set of six baseline scenarios, all with 500,000 exposed, with the number of predispensed MCMs varying from 0 to 5 million (in increments of 500,000) and used the MCM model to estimate the expected number of deaths for these 66 cases. We then changed one group (either lower or higher) or lowered p_M. After changing the parameter values, we evaluated the same 66 cases and used the MCM model to estimate the expected number of deaths and then divided the result by the expected number of deaths from the corresponding baseline scenario.

Summarizing by averaging the relative values over the 66 cases will provide an initial view of the results. The "worst case" disease progression increased the expected number of deaths by approximately 77%. The "best case" disease progression reduced the expected number of deaths by approximately 19%. Increasing the expected awareness times increased the expected number of deaths by approximately 19%. Decreasing the expected awareness times reduced the expected number of deaths by approximately 9%. Increasing the expected recovery times increased the expected number of deaths by approximately 7%. Decreasing the expected recovery times reduced the expected number of deaths by approximately 9%.

Reducing the local capabilities increased the expected number of deaths by approximately 18%. Increasing the local capabilities reduced the expected number of deaths by approximately 6%. Increasing the delays in response increased the expected number of deaths by approximately 12%. Reducing the delays in response reduced the expected number of deaths by approximately 7%. Reducing p_M to 50% increased the expected number of deaths by approximately 10%. Reducing the size of the push pack increased the expected number of deaths by approximately 1%. Increasing the size of the push pack reduced the expected number of deaths by approximately 1%.

Because the values of the disease progression parameters had the most impact on the expected number of deaths, we also estimated the sensitivity of the MCM model to these parameters, which were grouped into the following subsets: incubation time, progression time, and time to die. We increased and decreased the average times by 50%. For the time to die, we achieved this by multiplying the death rates by 2 (which reduced the average time to die by 50%) and by 2/3 (which increased the average time to die by 50%). We evaluated the same scenarios with the same number of predispensed MCMs.

The average relative values over the 66 cases are as follows: The shorter expected time to incubation increased the expected number of deaths by approximately 37%. The longer expected time to incubation decreased the expected number of deaths by approximately 14%.

The shorter expected time to become fulminant increased the expected number of deaths by approximately 21%. The longer average time to become fulminant decreased the expected number of deaths by approximately 11%.

The shorter expected time to die reduced the expected number of deaths by approximately 1%. The longer average time to die increased the expected number of deaths by approximately 1%. These small but counterintuitive results may result from the increased availability of resources such as ICU beds for treating others who are fulminant.

The impact of predispensing MCMs (the change in mortality due to predispensing) was affected by the parameter values. The "best case" disease progression most reduced the impact of predispensing MCMs across all six scenarios. Larger awareness rates, longer delays in response, fewer local capabilities, and the "worst case" disease progression increased the impact of predispensing MCMs. Figure 7.4 shows the

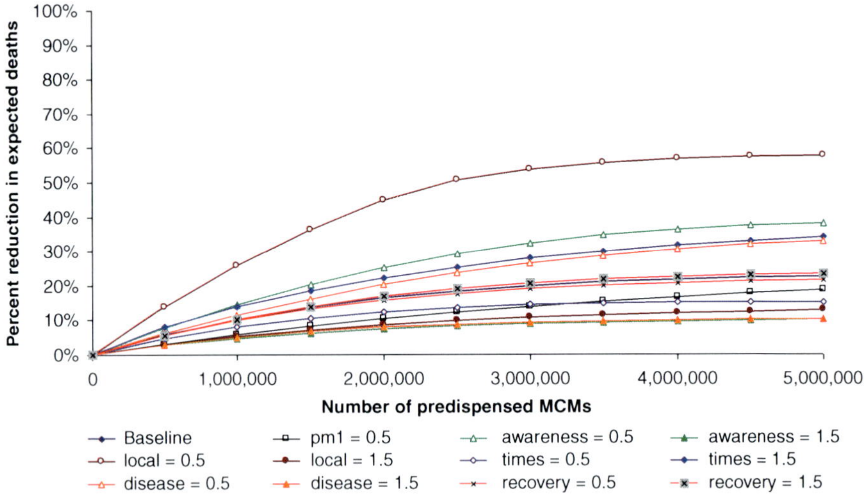

Fig. 7.4 Percent reduction in expected number of deaths as the number of predispensed MCMs increases, for the baseline and different settings of the parameter groups, in the scenario in which beta = 10 % and alpha = 90 %

percent reduction in the number of expected deaths (as the number of predispensed MCMs increases) for the various settings of the parameter values. The label "pm1 = 0.5" refers to reducing the parameter p_M to 50%. The label "awareness = 0.5" refers to reducing the "awareness" parameters by 50%; "awareness = 1.5" refers to increasing the "awareness" parameters by 50%. The labels for "local," "times," "disease," and "recovery" are used in the same way. (Note that, because they made a small change in the expected number of deaths, the changes to the settings of the size of the push pack are not included in this chart).

7.4 Discussion of MCM Model Results

The MCM model can estimate the number of deaths that result from an anthrax attack when a community predispenses MCMs and uses PODs to dispense prophylaxis to those without MCMs. The results show that, as more MCMs are predispensed, the expected number of deaths and the mortality rate decrease. The reduction in mortality rate is greater when the number of potential exposures is large. Essentially, predispensing MCMs counteracts the problems caused by the large number of potential exposures delaying the prophylaxis of those who were truly exposed. Thus, MCMs help both those who have them and those who don't. In addition, the MCM model can estimate the relative impact of increasing the local stockpile (instead of predispensing MCMs), but the results show that, in the scenarios considered, predispensing MCMs are much more effective when the mortality rate is high.

As more MCMs are predispensed, the adherence rate affects the mortality rate more significantly than the number exposed, the number of potential exposures, and the push pack delay. This indicates that predispensing MCMs should be accompanied by messages about the importance of adherence and techniques to avoid or reduce side effects that would discourage adherence (Werner and Deasy 2009).

The results also highlight the impact of the potential exposures. Public health officials should strive to reduce the number of persons not exposed who believe that they were exposed by determining the time and location of the attack as precisely as possible and by widely disseminating information about precisely who should receive prophylaxis (Brandeau et al. 2008).

The sensitivity results show that the disease progression has the largest impact on the expected number of deaths, especially if the disease progression is "worse" than the baseline. In that case, it is important to begin treatment as quickly as possible, so predispensing MCMs has more impact. Smaller awareness rates also have a large impact on the expected number of deaths because slowing awareness delays the administration of predispensed MCMs, so that predispensing MCMs has less impact. When awareness rates are larger, however, predispensing MCMs has more impact because they will be taken promptly after an attack. Changes in the local capabilities and the delays in response also affect the expected number of deaths. When local capabilities are fewer or the delays in response are longer, the effectiveness of the PODs is reduced, so predispensing MCMs has more impact. If p_M is reduced to 50%, the expected number of deaths increases because it essentially reduces the number of predispensed MCMs available. The impact is greatest in those cases when there are a larger number of potential exposures; in these cases predispensing MCMs can greatly reduce the expected number of deaths.

The numerical results discussed here are for some particular scenarios. The MCM model is capable of modeling many different scenarios by changing the parameters used in the model. While not discussed here, the MCM model can estimate the number of persons hospitalized and the total number of patient-days.

Because the MCM model is based on the POD model, it has many of the same limitations. Both models take the number exposed as an input (without modeling the release and dispersion of the anthrax spores). Both models make certain assumptions about the progression of anthrax, most particularly the assumption that the progression is not affected by age. Neither considers the impact of the disease on prophylaxis and treatment capacity. Both models assume that becoming aware of the attack is a simple diffusion process. In real communities, however, the spread of information may be quite different, which could affect the rate at which exposed persons and potential exposures seek and begin prophylaxis. Both models are deterministic compartmental models that estimate the expected outcome but give no insight into the distribution of possible outcomes for a given scenario. Houck (2011) considered a stochastic version of this model and found that the predictions of the models were very close and that the standard deviation of the number of deaths was small (about 1% of the mean).

The MCM model assumes that those who possess predispensed MCMs have the same chance to be exposed or to be potential exposures as those who do not have MCMs. Moreover, the model assumes that the progression of anthrax in exposed persons who take the antibiotics in the MCMs is the same as in exposed persons who receive prophylaxis from a POD. The compartments and flows in the model could be modified to represent scenarios in which these assumptions and others are replaced by other conditions.

7.5 Multiple City Resource Allocation Problem

Now we consider the problem of allocating a store of MCMs to multiple urban areas. Prepositioning MCMs in a city reduces the expected number of deaths from an anthrax attack in that city. The defender allocates MCMs before knowing which city the terrorist will attack. The terrorist (attacker) wishes to maximize expected number of deaths and will exploit any weaknesses in the defender's strategy. Thus, the defender, to minimize expected fatalities, must consider the attacker's decision. The approach presented here finds the optimal allocation.

The approach taken utilizes game theory, which can improve our understanding of "the nature of the key decisions that intelligent attackers and defenders must make" and emphasizes "that vulnerability and consequence are usually functions of the allocation decisions made by the players" (Cox 2009).

In particular, the problem is one of analyzing a leader–follower (or Stackelberg) game. The defender is the leader who must first allocate the MCMs, and the attacker is the follower who observes the defender's allocation before deciding which city to attack.

The mathematical model used to analyze the MCMs allocation problem has the following notation. There is a set of n potential targets (urban areas, or cities). The defender has a total budget of B MCMs available and will allocate these to the cities. Let $L_i(c_i)$ be the expected number of deaths in city i when c_i MCMs are prepositioned in that city. We assume that $L_i(c_i)$ is a continuous, monotonically decreasing function. Let P_i be the population of city i. This is the upper limit on c_i, and $L_i(c_i)$ reaches its minimum at this value. Because the $L_i(c_i)$ are monotonically decreasing, they can be inverted: $c_i = L_i^{-1}(y)$.

After observing the defender's allocation, the attacker wishes to maximize his expected utility, so he will attack the target that has the greatest value of $L_i(c_i)$. Let $h_i(c_1, \ldots, c_n) = 1$ if the terrorist will attack target i (that is, $L_i(c_i)$ is the maximum) and 0 otherwise.

The defender's objective is to minimize the total expected loss from the terrorist attacks:

$$\min_{c_1,\ldots,c_n} \sum_{i=1}^{n} h_i(c_1, \ldots, c_n) L_i(c_i)$$

subject to the budget constraint

$$\sum_{i=1}^{n} c_i \leq B.$$

Note that $h_i(c_1, \ldots, c_n)L_i(c_i) = \max_j L_j(c_j)$ for the target i that will be attacked and is 0 otherwise. Thus,

$$\sum_{i=1}^{n} h_i(c_1, \ldots, c_n)L_i(c_i) = \max_{j=1,\ldots,n} L_j(c_j).$$

In an optimal solution, the defender should invest resources in (allocate MCMs to) the cities in such a way that equalizes the expected number of deaths in the cities that receive MCMs (while the expected number of deaths in any cities without MCMs is even lower).

The range of expected number of deaths can be determined as follows:

$$L_{\text{max}} = \max_{i=1,\ldots,n} L_i(0),$$
$$L_{\text{min}} = \max_{i=1,\ldots,n} L_i(P_i).$$

It is not possible to reduce the expected number of deaths beyond L_{min}, so there is an upper limit on the number of MCMs that should be allocated, and, in some cases, there is no benefit to distributing any MCMs to cities that will have a low expected number of deaths (because they are otherwise well-prepared to respond to an anthrax attack).

Without loss of generality, renumber the cities so that $L_1(0) \geq L_2(0) \geq \cdots \geq L_n(0)$.

For each city i, if $L_i(0) \geq L_{\text{min}}$, let c_i^{max} be the value of c_i such that $L_i\left(c_i^{\text{max}}\right) = L_{\text{min}}$ (such a value must exist because $L_i(0) \geq L_{\text{min}} \geq L_i(P_i)$); otherwise, set $c_i^{\text{max}} = 0$. Then, the upper limit on the MCM allocation equals

$$B_{\text{max}} = \sum_{i=1}^{n} c_i^{\text{max}}$$

When $B = B_{\text{max}}$, the optimal allocation to city i is c_i^{max}.

If B (the total number of MCMs) is small, then the optimal solution allocates MCMs to only the cities with the most expected fatalities. As B (the total number of MCMs) increases, more cities will receive MCMs. Thus, it is valuable to determine the values of B at which additional cities are added to the set of those that receive MCMs. Let h be the number of cities with $c_i^{\text{max}} > 0$. (Note that these will be cities 1 to h).

Then, let c_{ij}^* be the value of c_i such that $L_i\left(c_{ij}^*\right) = L_j(0)$ for $i<j \le h$. Let $B_1 = 0$. Then, for $j = 2, \ldots, h$, define the breakpoints

$$B_j = \sum_{i=1}^{j-1} c_{ij}^*$$

If $B = B_j$, the optimal allocation is $c_i = c_{ij}^*$ for $i < j$ and $c_i = 0$ for $i \ge j$. The number of expected number of deaths equals $L_j(0)$. If $B \le B_j$, no MCMs are allocated to cities j to n. If $B > B_h$, then, in the optimal allocation, $c_i > 0$ for $i \le h$ and $c_i = 0$ for $i > h$.

Given a value of B in the range $[0, B_{\max}]$, the optimal allocation can be found as follows (1) let j^* be the largest value of j such that $B > B_j$; (2) find the value of y such that $\sum_{i=1}^{j^*} L_i^{-1}(y) = B$ (because this sum is a monotonically decreasing function of y, a bisection search or other similar technique can be used) and then set $c_i = L_i^{-1}(y)$ for $i = 1, \ldots, j^*$ and $c_i = 0$ for all other cities.

7.6 Resource Allocation Results

To illustrate this technique, we will consider the following scenario in which a store of MCMs will be prepositioned to the ten urban areas (in the USA) that have the highest expected annual terrorism losses (Willis et al. 2005). Table 7.4 lists the urban areas and their populations from the 2000 U.S. Census. In this example, all prepositioned MCMs are predispensed to the public.

We assume that the terrorist has enough anthrax to expose 500,000 individuals in any one of the cities. (We will also consider scenarios in which the number of exposures is 250,000 and 750,000, but we do not consider scenarios in which the terrorist simultaneously attacks more than one city because our results indicate that

Table 7.4 Ten urban areas and their populations

Urban area	Population (2000)
Los Angeles-Long Beach	9,519,338
New York	9,314,235
Chicago	8,272,768
Philadelphia, PA-NJ	5,100,931
Washington, DC-MD-VA-WV	4,923,153
Houston	4,177,646
Boston, MA-NH	3,406,829
Seattle–Bellevue–Everett	2,414,616
Newark	2,032,989
San Francisco	1,731,183

Table 7.5 Range of expected number of deaths for each urban areawhen 500,000 persons are exposed

Urban area	$L_i(P_i)$	$L_i(0)$	c_i^{max}
Los Angeles-Long Beach	41,396	130,375	9,427,066
New York	41,396	130,823	9,224,588
Chicago	41,397	133,423	8,196,345
Philadelphia, PA-NJ	41,400	147,310	5,063,803
Washington, DC-MD-VA-WV	41,401	148,567	4,888,162
Houston	41,402	154,859	4,151,522
Boston, MA-NH	41,405	163,825	3,389,652
Seattle–Bellevue–Everett	41,410	181,999	2,408,157
Newark	41,414	192,509	2,030,292
San Francisco	41,418	203,125	1,731,183
B_{max}			50,510,771

splitting the same total number of exposures among multiple cities is suboptimal for the attacker).

The scenario considered here is similar to the scenarios considered in Section 7.2 except for the following parameters. The number of potential exposures equals 25% of the number not exposed. The adherence rate is 90%. Intravenous antibiotics (IVs) for treatment, antibiotics for dispensing, and additional ventilators from the push pack will become available 16 hours after attack detection at $t = 64$ hours. A city's maximum prophylaxis dispensing capacity, which depends upon the population P_i, equals $P_i/1000$ persons per hour. The local stockpile has a one day dose of medication for every 100 persons and one day of IV antibiotics for every 10,000 persons. There is one ICU bed available for every 2,500 persons, and there is one respiratory technician available for every 25,000 persons.

The expected number of deaths in city i from an anthrax attack (when MCMs are predeployed and PODs are used after an attack) was estimated using the MCM model. In this scenario, each urban area was considered as one population, the number of predispensed MCMs was set to 0, 1%, 2%, . . ., 100% of the urban area's population, and the MCM model was used to estimate $L_i(c_i)$. For values of c_i other than those 101 values evaluated, we use a linear interpolation to approximate $L_i(c_i)$.

For this scenario (which we call Scenario 1), the expected number of deaths in each city can vary within the ranges shown in Table 7.5, which also shows the c_i^{max} for each city. Note that $B_{max} = 50{,}510{,}771$ MCMs. With no MCMs, San Francisco has the greatest number of expected number of deaths because, in the scenario considered here, it has fewer resources than the other urban areas. Therefore, it will be the first city to receive MCMs. Los Angeles will be the last.

Regardless of the number of MCMs distributed, some number of deaths is unavoidable. The unavoidable deaths result from the delays in detecting the attack and starting prophylaxis (during which time some exposed persons become very ill), the loss of MCMs among those who received them, and the imperfect adherence rate.

For any given value of B between 0 and 50,510,771, we can determine the optimal allocation to minimize the expected number of deaths using the procedure presented earlier. We also evaluated a simple allocation rule in which the number of

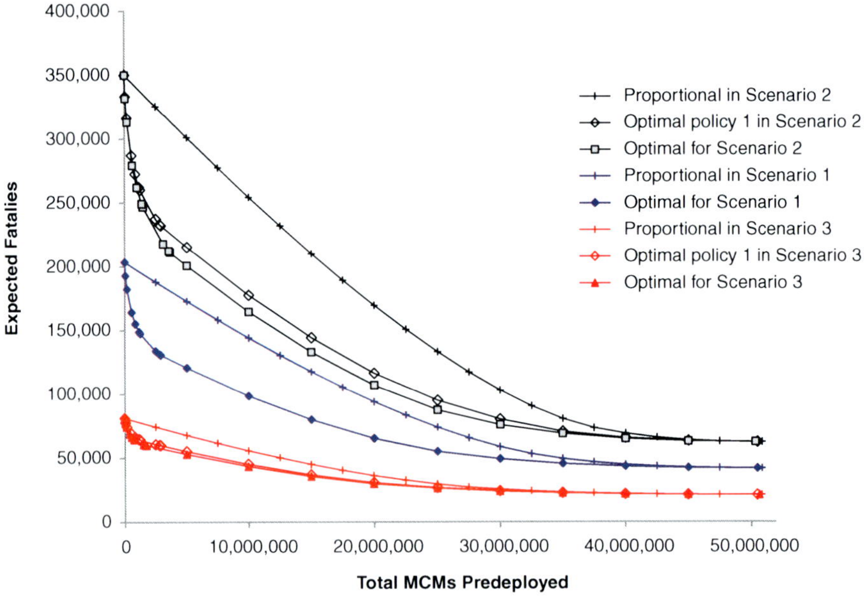

Fig. 7.5 Expected number of deaths over the range of B, optimal and proportional allocations, and three scenarios (1) 500,000 exposed, (2) 750,000 exposed, and (3) 250,000 exposed

MCMs allocated to each city is proportional to that city's population. Of course, when B is very large (approaching the total population of all ten urban areas), the optimal and proportional allocations are nearly the same and yield the same expected number of fatalities. When B is low, however, the allocations are very different, and the proportional allocation yields a higher expected number of fatalities, as shown in Fig. 7.5. Notice, in particular, how the expected number of deaths drops quickly as B increases when the optimal allocation is used.

We also considered the uncertainty in the attack scenario. This study investigated the uncertainty in the number of exposed, which could vary if the terrorist has more (or less) anthrax or if the conditions during the attack increase (or decrease) the number exposed. We let the number of exposed equal 750,000 (Scenario 2) and 250,000 (Scenario 3) and found, for various values of B, the optimal allocations for these new scenarios.

For any value of B, the allocation of MCMs that is optimal for Scenario 1 (500,000 exposed) is not optimal in Scenarios 2 and 3. We evaluated Scenario 1's optimal allocation and the proportional allocation in these new scenarios and compared the expected number of deaths to those that result from the optimal allocations for these scenarios, also shown in Fig. 7.5.

In these scenarios, the proportional allocation yields an expected number of deaths that is greater than the minimal expected number of deaths. The difference between the expected number of deaths with original optimal allocation and the scenario-specific optimal allocation is not as great. For instance, when

Table 7.6 Allocations of MCMs to each urban area under three different policies when $B = 20{,}000{,}000$

Urban area	Optimal for 750,000 exposed	Optimal for 500,000 exposed	Proportional allocation
Los Angeles-Long Beach	3,177,574	3,307,654	3,740,872
New York	3,128,173	3,249,777	3,660,271
Chicago	2,874,805	2,955,753	3,251,000
Philadelphia, PA-NJ	2,062,797	2,038,265	2,004,544
Washington, DC-MD-VA-WV	2,014,199	1,985,396	1,934,681
Houston	1,805,264	1,759,649	1,641,715
Boston, MA-NH	1,576,456	1,518,868	1,338,802
Seattle–Bellevue–Everett	1,251,882	1,189,857	948,886
Newark	1,113,225	1,053,957	798,916
San Francisco	995,625	940,824	680,313

$B = 20{,}000{,}000$ MCMs and the number exposed equals 750,000, the optimal policy allocates more MCMs to San Francisco than the proportional policy does (see Table 7.6). The expected number of deaths is 106,809 if the optimal allocation is selected, 116,150 (which is 9% greater) if Scenario 1's optimal allocation is selected, and 169,296 (58% greater) if the proportional allocation is selected. Thus, it appears that the original optimal allocation is robust with respect to the uncertainty in the number exposed.

7.7 Discussion of Resource Allocation Results

Clearly, if the defender has enough MCMs for everyone in every city, the allocation decision is trivial. When the number of MCMs is low, however, the allocation decision has a significant impact on the expected number of fatalities. Optimally allocating the MCMs is much better than a proportional allocation. Moreover, the optimal allocation for one scenario can be a very good allocation even if the scenario changes, which indicates that it is a robust solution.

These results also show that hedging (allocating resources to targets that are initially less attractive to the attacker) is optimal when there are sufficient resources. Previous work has shown that, in the optimal resource allocation, the most valuable target receives most of the resources when cost-effectiveness (the rate at which investments reduce the probability of a successful attack) is low, but, as cost-effectiveness increases, hedging becomes optimal, and more targets receive some resources for defense (Bier et al. 2008). In the context of MCMs allocation, cost-effectiveness is not directly relevant, but the total number of MCMs available for allocation does affect how many cities receive MCMs.

This study does not address the question of how many MCMs should be obtained, but the results seem to indicate that the marginal benefit of additional

MCMs is large if they are allocated optimally. As the number of MCMs available increases, the marginal benefit of additional MCMs decreases (this was seen also in the results shown earlier). A complete analysis of the appropriate investment in MCMs should also consider the financial, legal, regulatory, safety, and ethical considerations.

7.8 Conclusions

Public health officials who are considering whether and to what extent to preposition MCMs should consider how this option reduces the consequences of an anthrax attack. The MCM model described here enables that evaluation and should be valuable to these decision-makers.

In addition, the MCM model can be used as part of an approach for determining the allocation of a store of MCMs to multiple cities.

The decision to preposition MCMs depends upon many factors besides the expected impact on deaths. The cost of purchasing, predispensing, and replacing the MCMs may be substantial. The distribution should be fair, but issues of equity will appear if the number of MCMs available is not enough for everyone who wants them. The risks of individuals using the MCMs for unauthorized, possibly dangerous, uses must be assessed as well. Legal and regulatory issues must also be considered. A complete analysis of this decision is the focus of future work.

This chapter discussed the problem of allocating a store of MCMs to multiple cities. A game theory-based approach is adopted, and the attacker's objective is used to define the objective function that the defender needs to optimize. In particular, the objective is to minimize the maximum expected number of deaths.

When the total number of MCMs is low, the optimal solution allocates MCMs to a small number of cities that have the highest expected number of deaths. When more MCMs are available, all of the cities receive some, but the optimal allocation is not proportional to the cities' populations. Based on this analysis, finding the optimal solution is not difficult. An illustrative example was used to demonstrate the essential characteristics of the problem.

This study considered only the allocation of MCMs that will be predispensed to the general population. Predispensing MCMs to first responders and other personnel who are essential to continuity of operations will reduce the number of MCMs available to the general population. In general, the allocation of scarce resources to different urban areas and to groups within an urban area must be considered within a framework of ethical guidelines that emphasize the relevant moral principles (Kinlaw et al. 2009).

Acknowledgments The author received no external funding for this study. The author gratefully acknowledges the contributions of Michelle L. Houck, who constructed mathematical models and collected numerical results using these models to predict the impact of predispensing medical countermeasures.

References

Bicknell WJ (2003) How best to fight against bioterrorism. http://www.cato.org/pub_display.php?pub_id=3189. Accessed 8 June 2011

Bier VM, Haphuriwat N, Menoyo J et al (2008) Optimal resource allocation for defense of targets based on differing measures of attractiveness. Risk Anal 28:763–770

Brandeau ML, Zaric GS, Freiesleben J, Edwards FL, Bravata DM (2008) An ounce of prevention is worth a pound of cure: improving communication to reduce mortality during bioterrorism responses. Am J Disaster Med 3:65–78

Centers for Disease Control and Prevention (2007) Emergency MedKit evaluation study summary. http://www.bt.cdc.gov/agent/anthrax/prep/pdf/medkit-evaluation-summary-2007.pdf. Accessed 7 June 2011

Centers for Disease Control and Prevention (2011) Cities readiness initiative. http://www.bt.cdc.gov/cri/. Accessed 7 June 2011

Cox LA (2009) Game theory and risk analysis. Risk Anal 29:1062–1068

Graham B, Talent J, Allison G, Cleveland R, Rademaker S, Roemer T, Sherman W, Sokolski H (2008) World at risk: the report of the commission on the prevention of WMD proliferation and terrorism. Vintage Books, New York

Holty JE, Bravata DM, Liu H, Olshen RA, McDonald K, Owens DK (2006) Systematic review: a century of inhalational anthrax cases from 1900 to 2005. Ann Intern Med 144:270–280

Houck ML (2011) Compartmental and simulation models for evaluating MedKit prepositioning strategies for anthrax attack response. University of Maryland, College Park, MD

Houck ML, Herrmann JW (2011) Predicting the impact of placing pre-event pharmaceuticals for anthrax. Technical report. Institute for Systems Research, University of Maryland, College Park, MD

Institute of Medicine (2008) Dispensing medical countermeasures for public health emergencies. Workshop summary. Institute of Medicine Forum on Medical and Public Health Preparedness for Catastrophic Events, National Academies Press, Washington, DC

IOM (Institute of Medicine) (2012) Prepositioning antibiotics for anthrax. The National Academies Press, Washington, DC

Jacquez JA (1996) Compartmental analysis in biology and medicine, 3rd edn. BioMedware, Ann Arbor, MI

Kinlaw K, Barrett DH, Levine RJ (2009) Ethical guidelines in pandemic influenza: recommendations of the Ethic Subcommittee of the Advisory Committee of the Director, Centers for Disease Control and Prevention. Disaster Med Public Health Prep 3:S185–S192

National Biodefense Science Board (2008) Excerpts from the summary report of the National Biodefense Science Board. http://www.phe.gov/Preparedness/legal/boards/nbsb/Documents/nbsb-excrpt-pp-080618.pdf. Accessed 6 July 2011

Oren M (2009) Biological agents and terror medicine. In: Shapira SC, Hammond JS, Cole LA (eds) Essentials of terror medicine. Springer, New York

Richter A, Khan S (2009) Pilot model: judging alternate modes of dispensing prophylaxis in Los Angeles County. Interfaces 39:228–240

Shepard CW, Soriano-Gabarro M, Zell ER et al (2002) Antimicrobial postexposure prophylaxis for anthrax: adverse events and adherence. Emerg Infect Dis 8:1124–1132

Stern PC, Fineberg HV (eds) (1996) Understanding risk: informing decisions in a democratic society. National Academy Press, Washington, DC

Troy T (2010) Preparing for bioterrorism. Weekly Standard, 23 Feb 2010. http://www.hudson.org/index.cfm?fuseaction=publication_details&id=6785. Accessed 8 June 2011

Walter GG, Contreras M (1999) Compartmental modeling with networks. Birkhäuser, Boston

Wein LM, Craft DL, Kaplan EH (2003) Emergency response to an anthrax attack. Proc Natl Acad Sci USA 100:4346–4351

Werner K, Deasy J (2009) Acute respiratory tract infections: when are antibiotics indicated? J Am Acad Phys Assist 22:22–26
Willis HH, Morral AR, Kelly TK et al (2005) Estimating terrorism risk. RAND Corporation, Santa Monica. http://www.rand.org/pubs/monographs/2005/RANDMG388.pdf
Zaric GS, Bravata DM, Holty JC, McDonald KM, Owens DK, Brandeau ML (2008) Modeling the logistics of response to anthrax bioterrorism. Med Decis Making 28:332–350

Chapter 8
Service Networks for Public Health and Medical Preparedness: Medical Countermeasures Dispensing and Large-Scale Disaster Relief Efforts*

Eva K. Lee, Ferdinand Pietz, and Bernard Benecke

Abstract A catastrophic health event, such as a terrorist attack with a biological agent, a naturally occurring pandemic, or a calamitous meteorological or geological event, could cause tens or hundreds of thousands of casualties, weaken the economy, damage public morale and confidence, create panic and civil unrest, and threaten national security. It is therefore critical to establish a strategic vision that will enable a level of public health and medical preparedness sufficient to address a range of possible disasters. Planning for a catastrophe involving a disease outbreak or mass casualties is an ongoing challenge for first responders and emergency managers. They must make critical decisions on treatment distribution points, staffing levels, impacted populations and potential impact in a compressed window of time when seconds could mean life or death. Some of the key areas of public health and medical preparedness include medical surge, population protection, communication infrastructure, and emergency evacuation. This chapter highlights our own experience on projects with the Centers for Disease Control and Prevention and various public health jurisdictions in emergency response and medical preparedness for mass dispensing for disease prevention and treatment and large-scale disaster relief efforts.

---

*The findings and conclusions in this report are those of the authors and do not necessarily represent the official position of the Centers for Disease Control and Prevention.

E.K. Lee (✉)
Center for Operations Research in Medicine and HealthCare, School of Industrial and Systems Engineering, NSF I/UCRC Center for Health Organization Transformation, Georgia Institute of Technology, 755 Ferst Dr. NW, Atlanta, GA 30332-0205, USA
e-mail: evakylee@isye.gatech.edu

F. Pietz
Strategic National Stockpile, Office for Public Health Preparedness Response, Centers for Disease Control and Prevention, Atlanta, GA, USA

B. Benecke
Global Disease Detection and Emergency Response, Center for Global Health, Centers for Disease Control and Prevention, Atlanta, GA, USA

J.W. Herrmann (ed.), *Handbook of Operations Research for Homeland Security*, International Series in Operations Research & Management Science 183, DOI 10.1007/978-1-4614-5278-2_8,

8.1 Introduction

A catastrophic health event, such as a terrorist attack with a biological agent, a naturally occurring pandemic, or a calamitous meteorological or geological event, could cause tens or hundreds of thousands of casualties or more, weaken the economy, damage public morale and confidence, create panic and civil unrest, and threaten national security. It is therefore critical to establish a strategic vision that will enable a level of public health and medical preparedness sufficient to address a range of possible disasters. Although present public health and medical preparedness plans incorporate the concept of "surging" existing medical and public health capabilities in response to an event that threatens a large number of lives, the assumption that conventional public health and medical systems can function effectively in catastrophic health events has proven to be incorrect in real-world situations. Therefore, it is necessary to transform the approach to health care in the context of a catastrophic health event in order to enable public health and medical systems to respond effectively to a broad range of incidents.

According to the Homeland Security Presidential Directive 21, public health and medical preparedness refers to "the existence of plans, procedures, policies, training, and equipment necessary to maximize the ability to prevent, respond to, and recover from major events, including efforts that result in the capability to render an appropriate public health and medical response that will mitigate the effects of illness and injury, limit morbidity and mortality to the maximum extent possible, and sustain societal, economic, and political infrastructure."

Planning for a catastrophe involving a disease outbreak or mass casualties is an ongoing challenge for first responders and emergency managers. They must make critical decisions on treatment distribution points, staffing levels, impacted populations and potential impact in a compressed window of time when seconds could mean life or death. Although extensive resources have been devoted to planning for a worse-case scenario on the local, regional, and national scale, the US Government Accountability Office (GAO) found that gaps still exist. While many states have made progress in planning for mass casualty events, many noted continued concerns related to maintaining adequate staffing levels and accessing other resources necessary to effectively respond.

This chapter highlights our own experience on projects with the Centers for Disease Control and Prevention (CDC) and various public health jurisdictions in emergency response and medical preparedness for mass dispensing for disease prevention and treatment and large-scale disaster relief efforts. Specifically, Section 8.2 provides a brief introduction and motivation for medical countermeasure dispensing. Section 8.3 offers a systems view of mass dispensing operations.

In Section 8.3.1, we describe various modes of dispensing and POD placement. This is followed by resource allocation and POD layout design in Sect. 8.3.2. Section 8.3.3 highlights the importance of disease propagation analysis and design of mitigation strategies. Section 8.3.4 describes supply chain management that includes demand, supply, fulfillment, and partnership. In particular, we briefly explain the mission and responsibility of CDC's Strategic National Stockpile. Section 8.3.5 discusses communication and public information. This entails the communication infrastructure (hardware tools and software programs) that supports emergency operations, as well as public information and risk communication. Section 8.3.6 offers insights on lessons learned from flu mass vaccination and anthrax drill exercises, and shares our knowledge on continued challenges and the need for multi-layer protection.

Section 8.4 summarizes large-scale disaster relief efforts, including the establishment of a network of service constructs for food, medical needs, and shelters; staffing and resource constraints; susceptibility of displaced population to infectious disease outbreaks; and supply chain management. Coordination among different stakeholders; risk and uncertainty that are faced by on-the-ground rescue and relief workers; and media exposure are also discussed.

In Sect. 8.5, we summarize some methodologies that are commonly employed in emergency responses. These techniques include optimization, stochastic processes, information technology, and an integrated framework of decision systems. The chapter concludes with some challenges for future research.

8.2 Population Protection: Medical Countermeasures Dispensing and Large-scale Disaster Relief Efforts

Public health emergencies, such as bioterrorist attacks or pandemics, demand fast, efficient, large-scale dispensing of critical medical countermeasures (i.e., vaccines, drugs, and therapeutics). Such dispensing is complex and requires careful planning and coordination from multiple federal, state, and local agencies and the potential involvement of the private sector. Dispensing medications quickly (within 48 h for anthrax prophylaxis) to large population centers (with tens of thousands or even millions of people) is urgent; moreover, the multi-faceted nature of dispensing (e.g., sending federal stockpiles to local points-of-dispensing (PODs), coordination at the local level to manage the transportation of citizens to PODs, and the POD operations) makes the process highly unpredictable. Thus, emergency managers and public health administrators must be able to quickly investigate alternative response strategies as an emergency unfolds.

The focus of this chapter is on *mass dispensing* of medical countermeasures for protection of the general population. Other issues pertinent to large-scale disaster relief efforts will also be highlighted. Much of the writing below are excerpts from our recent work with the Centers for Disease Control and Prevention on modeling and optimizing public health emergency response infrastructure and the development of a large-scale simulation-optimization decision support system, RealOpt (Lee et al. 2006a, b, 2009a, b); (Lee 2008).

Large-scale public health emergencies may involve thousands of sick or injured people who will require various levels of medical care, ranging from patient evacuation (Lee et al. 2011c), hospital care, and sustainable and potentially long-term health-recovery procedures. Thus, such emergencies present a daunting set of challenges, including the surge capacity and flexibility of our existing medical systems (Lee et al. 2011a, c), federal and state emergency capacity for rapid medical dispatching, and the resolve and resilience of health-care workers and emergency responders to perform under critical timelines and exceedingly stressful conditions.

In the wake of the 2001 anthrax attacks, the Department of Health and Human Services (HHS) increased its order for smallpox vaccine, accelerated production, and began working to develop a detailed plan for the public health response to an outbreak of smallpox. By January 2003, the USA had sufficient quantities of the vaccine for every person in the country in an emergency situation (Gerberding 2003). Subsequently, HHS required each state to submit a mass vaccination plan for administering smallpox vaccine. Further, states are charged with developing city-readiness programs that deal with establishing regional treatment and dispensing centers, and developing procedures, policies, and a planning framework for efficient allocation of staff and resources in response to these events.

The importance of such population protection has been carefully studied for human, social, and economic benefits. Kaplan et al. (2002) argued that immediate mass vaccination after a smallpox bioterrorist attack would result in fewer deaths and faster eradication of the potential epidemic; Wein et al. (2003) concluded that immediate and aggressive dispersion of oral antibiotics and the full use of available resources (local nonemergency care workers, federal and military resources, and nationwide medical volunteers) are extremely important.

8.3 A Systems View of Mass Dispensing Operations: Integrating all the Elements

Public health and medical preparedness involves three phases: (1) preparedness and prevention, (2) detection and response, and (3) recovery and mitigation.

Modeling and optimizing public health infrastructure involve elements of resource allocation under risk, uncertainty, and time pressure; large-scale supply chain management; transportation and operational logistics; and medical treatment and population protection. The operations must be supported by an effective communication infrastructure (Lee et al. 2011b). There is a necessity for vertical and horizontal integration and communication, where federal, state, local, tribal, territorial, private, and business stakeholders work toward a common goal of a resilient public health system. The infrastructure must be *flexible, scalable, sustainable, and elastic* to support an effective and timely response, and to mount rapid recovery and mitigation operations (Lee 2009; Lee et al. 2009a, b).

The global integration plan for population protection in the event of medical and emergency response includes multiple interconnected components: strategic planning; management command and control with key leaders working cohesively together as a team; requesting supplies and equipment from Strategic National Stockpiles; tactical communication and information technology; public information and risk communication; security; regional and local distribution sites (for receiving, staging, and storing of supplies, as well as transportation and routing); inventory control and management; distribution—supply and re-supply; dispensing; treatment centers; and planning, training and evaluation (Lee 2009). The ultimate goal is to *dispense* the medical countermeasures to the affected regional population in a timely manner.

8.3.1 Mode of Dispensing and POD Placement

Mass dispensing requires the rapid establishment of a network of dispensing sites and health facilities that are *flexible*, *scalable*, and *sustainable* for medical prophylaxis and treatment of the general population. Moreover, each POD must be capable of serving the affected local population within a specified short time frame. Clearly, for very large-scale dispensing, the sophisticated logistical expertise needed to deal with the complexities of selecting an adequate number of strategically well-placed POD locations, and of designing and staffing each POD, is beyond the capability of any human planner or public health administrator. The limited availability of trained critical staff, such as public health professionals, further compounds the inherent complexities.

The CDC and public health administrators work closely with one another to prepare for and document the steps required to administer medication in the event that mass dispensing is needed. The goal and objectives of a dispensing facility, *point-of-dispensing (POD)*, are to deliver appropriate emergency services (e.g., vaccine, medical service, and education/training) to high-risk populations in an orderly, expeditious and safe manner. Within the POD facility, the potential tasks and objectives may include

1. Assess health status of clients.
2. Assess eligibility of clients to receive service.
3. Assess implications of each case and refer case for further investigation if necessary.
4. Counsel clients regarding service and associated risks.
5. Administer service.
6. Educate regarding adverse events.
7. Document services.
8. Monitor vaccine take rates.
9. Monitor adverse reactions.
10. Monitor development of disease.

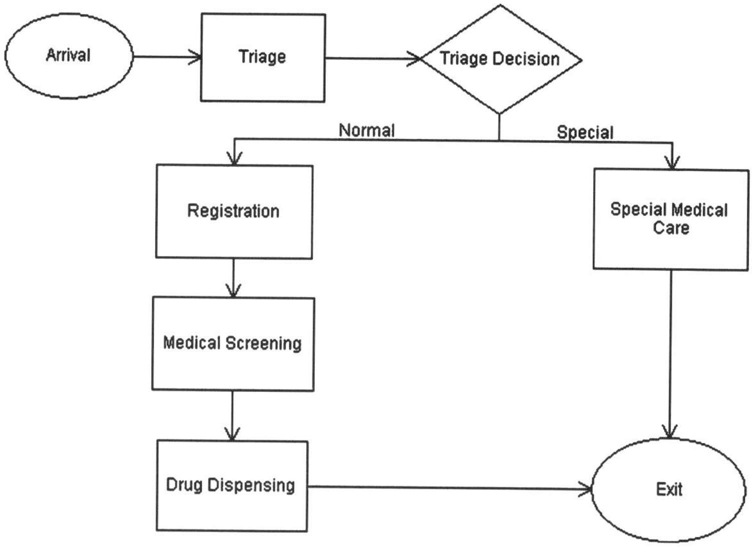

Fig. 8.1 The flowchart shows a POD that was set up in a national drill exercise to dispense anthrax antibiotics

8.3.1.1 Mode of Dispensing

The *key* to mass dispensing is to protect the general population efficiently and effectively under time pressure. For example, in an anthrax attack, the goal is for citizens to receive antibiotic prophylaxis within 48 h of the determination that an attack has occurred, as the mortality rate for persons demonstrating symptoms of inhalation anthrax is extremely high (Lawler and Mecher 2007). Thus, it is recognized that multiple dispensing modalities often must be employed in order to *serve (cover)* the entire regional population. For example, special dispensing services will be utilized to serve homebound, disabled, and special need populations. In some instances, it is unreasonable to expect residents to travel to a designated POD facility. For example, nursing homes, assisted living facilities, homeless shelters, hospitals, and prisons house many residents for whom it would be inconvenient or inadvisable to travel to a public dispensing facility. Moreover, in many of these instances, there are already medical personnel on site who can assist in the dispensing process. In such cases, it would be more efficient to set up a *closed* POD inside these locations for dispensing, or to have medication dispensed by a mobile POD facility near the site. In this case, the POD is *closed* as it provides services only to the residents on-site, and is not open to walk-ins from the general public. Corporate offices that staff a large number of employees could be served in a similar manner. Once these sites receive prophylactic supplies, they could set up a closed POD within their building, with their own health-care staff and volunteers, or with public health staff supplemented by the state. Several factors suggest that such closed PODs will have fewer security concerns and will be easier to manage than public PODs. These factors include familiarity with the environment and people (e.g., fellow residents/employees), existing security measures including established checkpoints and previously authenticated identification badges with photo and/or biometric markers, and less stress than having to commute to a public POD.

Airports and hotels, where a large number of nonresident travelers can be found, are also candidates for setting up PODs to service specific vulnerable populations. Universities can use their own health facilities (and if necessary, additional mobile on-campus PODs provided by the state) to provide prophylaxis to on-campus students, staff and faculty. Clearly, if large employers and medical facilities provide prophylaxis to their own employees, families, and patients, it will eliminate a high percentage of the population (may be as high as 40% in some large cities) from visiting public PODs, thus reducing the load on those facilities.

Public PODs are *open* facilities that are setup to serve the general public.

In our work with CDC (Lee et al. 2006a, b, 2009a, b), public PODs have generally been described as being setup inside existing facilities, or in outdoor tents, with areas set aside for various activities in the dispensing process, including assembly/intake, triage, orientation, registration, screening, service, education and discharge. Public PODs can be mobile or stationary, and in the latter case, they can be setup as *facility-based* or *drive-through*.

A facility-based POD operates within physical locations, such as buildings, warehouses, open fields, or large parking lots. Citizens are asked to arrive at the POD location and then *walk through* the POD to receive their medication or other treatment. Facility-based PODS may be scaled to operate within a setting as large as a professional sports stadium or as small as a volunteer fire house within a rural community.

These walk-through PODs are suitable for relatively large capacity facilities, where the possibility of traffic jams preclude the use of the drive-throughs. Because parking is typically limited, individuals may be directed to arrive at designated points, and they are bused to the POD. Examples of pickup points include bus stations, subway stations, and parking lots of large shopping malls, where sufficient parking is available.

High schools are often selected as potential POD locations. The fact that they are government-owned makes logistics easier. Other considerations include existence of offices, computer communications, cafeterias, storage, etc. Shopping malls, churches, and stadiums are also suitable, and in some cases, PODs are set up outdoors using tents and temporary constructs.

Drive-through PODs are suitable to serve a spread-out population. Ideally, government-owned properties are preferred over privately owned ones for logistics reasons. Locations for setting up drive-through facilities should have enough space for multiple dispensing lanes; surrounding access roads; and room for command tent, employee rest area, and medication storage.

8.3.1.2 POD (Facility) Placement

Facility location problems are classic optimization problems and have been a critical element in strategic planning for a wide range of private and public organizations. The earliest facility location problems incorporating emergency response relate to location of emergency facilities (Swain et al. 1971; Larson 1975; Aly and White 1978). Chaiken and Larson (1972) provide a survey on urban emergency unit allocation. Some researchers (Hogan and Revelle 1986; Pirkul and Schilling 1988; Narasimhan et al. 1992) explicitly take the need for backup facilities (in case the main facility is overloaded) into account. More recently, Jia et al. (2002) provides a modeling framework for location of medical services for large-scale emergencies. Berman and Gavious (2007) take a game theoretic approach toward location of terror response facilities. Church et al. (2004) study the problem of identifying and protecting critical infrastructure. A comprehensive review of facility location research can be found in Brandeau and Chiu (1989), where the authors present a survey of over 50 representative problems in location research. Most of the problems reviewed have been formulated as optimization problems. Owen and Daskin (1998) provide another review of the strategic facility location problem. They consider a wide range of model formulations across numerous industries, including stochastic formulations, and discuss solution approaches.

While many of the facility location problems involve permanent construction and establishment of service facilities, facility location problem for mass dispensing concerns the placement of PODs (mobile or stationary) in existing facilities (e.g., high schools, stadiums, shopping malls, open parking lots) in a region to provide the necessary services to the population within a designated period of time.

Various objectives can be incorporated within the models. In the event of catastrophic incidents, it is critical that PODs are strategically located so as to allow easy access by the affected public. Hence, minimizing transportation time can be one critical objective. Further, the setup and operating costs of PODs cannot be neglected. A POD must be accessible by service workers, it should include a good communication infrastructure, it must be easily protected by law-enforcement personnel, and the facility must be capable of handling a large flow of people. Physical constraints on the facility must be modeled properly, e.g., capacity of a facility cannot be violated (e.g., POD parking capacity is limited, and fire codes limit the number of individuals who can be inside a facility simultaneously).

For operational purposes, there is also a desire to ensure that the number of PODs in each jurisdiction is at least two. This is due to the concern that if a catastrophic event at one site necessitates shutting down of a POD, emergency dispensing can still be carried out in the remaining location. In this case, the response manager can re-route populations to the remaining site, while re-establishing a new POD, if deemed necessary.

The problem of modeling POD placement involves the traditional facility location issues, as well as the incorporation of spatial, geographical, and demographic information. In our work with CDC, the problem is modeled using a two-stage integer programming approach. For a large metropolitan area with a population over five million, such a POD placement problem can result in optimization (integer programming) instances involving millions of variables and constraints for each jurisdiction that consists of hundreds of thousands of people in the region. The challenge is to determine the tradeoffs between the quality of solution, the practicality of planning, and real-time optimization. Exact algorithms and heuristics must be developed and advanced to address such computational challenges (Lee et al. 2009a, b, 2013).

8.3.2 Resource Allocation and POD Layout Design

8.3.2.1 Resource Allocation

Mason and Washington (2003) at CDC investigated optimal staffing arrangements for dispensing sites in the face of limited resources via a simulation/optimization system "Maxi-Vac" that they developed. Their study offered insight on the practicality of such a system as a planning tool for emergency situations, but revealed critical bottlenecks between the commercial simulation software and the optimization software: over 10 h were needed to obtain a usable feasible solution in each scenario with about 25–30 staff. This initiated our collaboration with CDC and the development of RealOpt, a large-scale real-time simulation-optimization decision support system (Lee et al. 2006a, b, 2009a, b, c, 2013).

Given a staff assignment (obtained from an initial optimization step), and input of service distributions at each station, we can model and simulate the movement of individuals inside a POD. The simulation output is a set of parameters (including statistics of average flow time, queue length, wait time, utilization rate, etc.) that enable evaluation of the objective function being optimized (e.g., the resulting throughput).

The optimization of labor resources involves placement of staff at various stations in the POD to maximize throughput or minimize the staffing needs to satisfy a preset throughput population. The cost at each station depends on the type and number of workers who are assigned to that station and have the required skills and on the average wait time, queue length, and utilization rate of the station. The total system cost depends on the cost at each station, and on system parameters such as cycle time and throughput. These cost functions are not necessarily expressible in closed form.

Constraints in the model include maximum limits on average wait time and queue length, range of utilization desired at each station, and upper and lower bounds on the number of workers with the required skills who are needed to perform various tasks at the POD. Constraining the average cycle time to be less than a prespecified upper bound is critical for emergency response, because individuals must move through the system as quickly as possible to facilitate crowd control, reduce sources of human frustration and potential disorderly

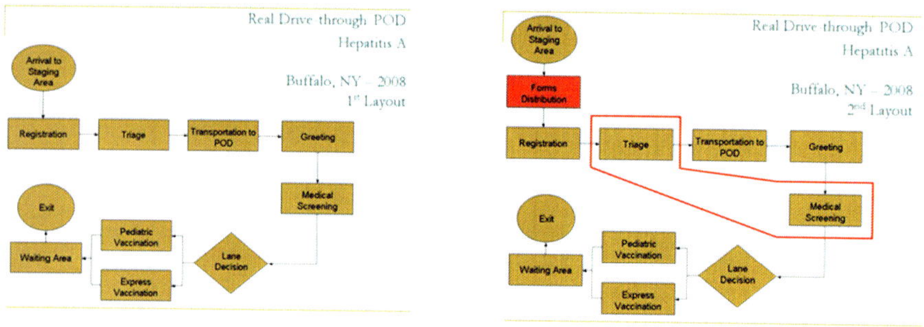

Fig. 8.2 Two POD layouts used during an actual 2008 Hepatitis vaccination event

outbursts, and reduce the potential spread of disease or contamination. The resulting nonlinear mixed-integer program poses unique challenges for existing optimization engines (Lee et al. 2006a, b, 2009a, b, c, 2013).

8.3.2.2 POD Layout Design

Designing the appropriate POD for various medical dispensing is critical. Further, POD layout will affect the overall staffing and efficiency of the dispensing operations.

Figure 8.2 contrasts two POD designs that were employed in a Hepatitis A booster shot event for 10,000 citizens (The Buffalo News 2008). The left shows the drive-through POD design used in the morning, and the right shows the re-design in the afternoon after real-time reconfiguration (based on service times collected on site) was performed. The re-design offers 10% improved throughput, 18% improved utilization, and a range of 10–85% reduction in wait time and queue length at various stations. *This illustrates the paramount importance of POD design to any emergency operation where resources are scare, time is precious, and there is a large affected population to serve.*

8.3.3 Disease-Propagation Analysis: Mitigation Strategies and Choice of Dispensing Modalities

Large-scale dispensing clinics could facilitate the spread of disease because of their high-volume population flow. The field of dynamical systems (mostly differential equation systems) provides the principle methods of modeling in classical mathematical epidemiology (Anderson et al. 1992; Diekmann and Heesterbeek 2000). Despite their simplicity when compared to recent complex simulation studies (Ferguson et al. 2005, 2006; Longini et al. 2005; Germann et al. 2006), these methods have helped generate functional insights, such as the transmission threshold for the start of an epidemic and the vaccination threshold for containment of an outbreak. As modelers attempt to incorporate more realistic dynamics into their models (such as stochasticity, nonexponential waiting times, sample-path-dependent events, and demographical and geographical data), more flexible tools, such as individual-based stochastic simulations, are preferable. Although simulation is a powerful approach, it is less mathematically tractable (i.e., it requires intensive computing time) than the classical methods.

The rapid and large-scale simulator in RealOpt opens up an opportunity to explore disease-propagation studies in which stochasticity of systems can be incorporated readily. It includes a disease-propagation module that aids users in understanding facility design and flow strategies that mitigate the spread of disease. The module incorporates the standard four-stage SEIR (susceptible, exposed, infectious, and recovered) model (Kermack and McKendrick 1991), and a novel six-stage SEPAIR model to capture the disease development (i.e., asymptomatic or symptomatic). By distinguishing the symptomatic stage from the asymptomatic stage, this model allows one to examine the effect of triage accuracy in POD facility design.

Lee et al. (2009b, 2010a) give a detailed theoretical and computational analysis of disease propagation and strategies for mitigation during biological or pandemic outbreaks and mass dispensing. In addition to the incorporation of stochasticity of client arrival and service distribution into the model, it also accommodates the following factors.

- The clinic model can be represented as an n-server system with queuing; transmission can occur between clients or between clients and staff. (In a real emergency, staff members will be given medical countermeasures to protect them from the disease prior to their assignment to POD services. However, a medical countermeasure does not provide 100% protection; each staff member still has a small probability of being infected by clients).
- The intra-clinic infectivity between clients and staff can vary.
- If symptomatic individuals are not triaged out properly during the initial screening, they could infect other people inside the POD. The system allows users to observe the effect of triage and screening errors, determine improved strategies for triage and screening, and establish guidelines for mitigating the spread of disease because of such errors.
- Inhomogeneous mixing within the community is possible.
- The infectious, asymptomatic, and symptomatic individuals can infect at various rates.

Figure 8.3 contrasts the triage accuracy with respect to the symptomatic proportion, when simple mass-action incidence infection is considered (Lee et al. 2009b, 2010a). This analysis assesses errors in triage and their infection consequences. It provides estimates for POD planners and epidemiologists to help determine the level and expertise of triage that should be in place with respect to the transmission coefficient. It also allows for scenario-based comparison of effective POD design.

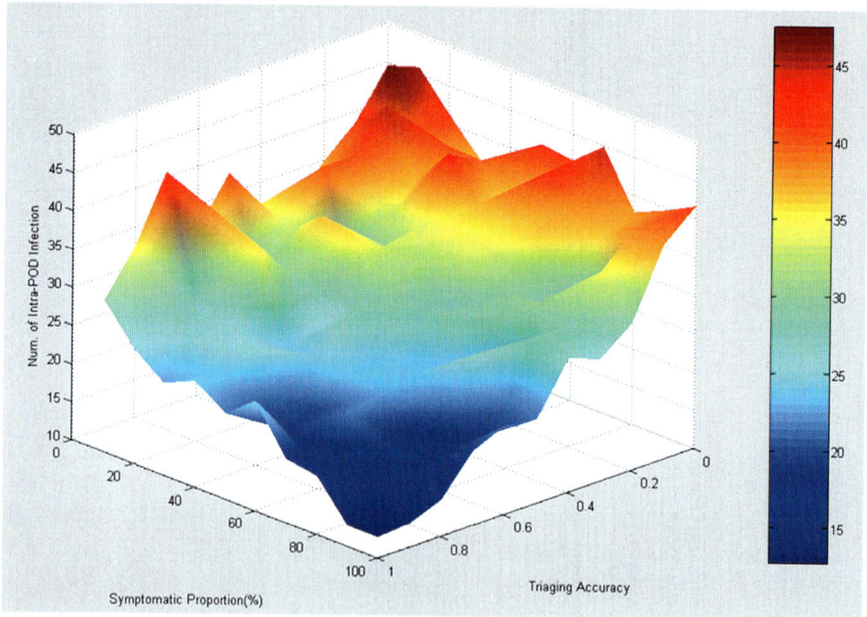

Fig. 8.3 The graph shows the triage accuracy versus symptomatic proportion and the importance of using the SEPAIR six-stage propagation model, because it allows us to examine the effect of implementing triage accuracy. The graph shows the number of intra-POD infections (*vertical axis*) under different triage accuracy and symptomatic proportions (*horizontal axes*). The throughput is 36,000 over a period of 36 h. The contact number is 193 (for outer-POD disease propagation), and the transmission coefficient is $0.18\mathrm{E}^{-5}$/min. The incoming percentage for susceptible is 95 %, and for infectious is 5 %. The mean dwell time is 1 day for both exposed and infectious and 3 days for asymptomatic and symptomatic

Such analyses may influence the selection of dispensing modalities. Specifically, over the past few years, we have observed more use of drive-through PODs for infectious disease prophylaxis dispensing (e.g., seasonal flu vaccination for communities and the H1N1 mass vaccination in 2009 and early 2010). In January 2008, a Hepatitis A confirmation of a grocery worker triggered the prophylactic vaccination of 10,000 residents in Erie County in New York who were potentially exposed to the disease, costing the county's public health agency at least $500,000. The health department dispensed the first vaccination in February when it set up a stationary clinic (walk-through POD). Because of the medical logistics and infectious nature of the disease, some people had to wait for hours in frigid temperatures. In September 2008, the health department used a Hepatitis A follow-up drive-through POD to provide the required second shot of vaccination. The POD was also the first test of the county's "drive-through" plan. The drive-through process is quick, efficient, and convenient, and minimizes the potential of intra-infectivity (The Buffalo News 2008).

8.3.4 Supply Chain Management: Demand, Supply, Fulfillment, and Partnership

Although supply chain management during a disaster response mirrors that of business supply chain management, damaged or destroyed infrastructure forces the use of ad hoc solutions that limit the effectiveness and efficiency of the operation.

Demand management can be extremely difficult due to the fluidity of the population and the potential collapse of the supporting infrastructure. Problems vary depending on the nature of the disaster. In the event of an infectious disease outbreak, as in the 2009 H1N1 event, when there is not a sufficient supply of medical countermeasures for the affected populations, predicting and allocating the proper distribution across the nation becomes critical, but demand is highly stochastic and uncertain (Lee et al. 2010b). Decisions can have a major impact on overall infectivity and mortality rates. In the same manner, demand within an earthquake zone fluctuates rapidly due to the movement of people fleeing from one site to another. And the time it takes to implement an effective response can be critical to the survivors of the affected population.

Within the USA, an act of terrorism (or a large-scale natural disaster) targeting the US civilian population will require rapid access to large quantities of pharmaceuticals and medical supplies. Such quantities may not be readily available unless special stockpiles are created. No one can anticipate exactly where a terrorist will strike and few state or local governments have the resources to create sufficient stockpiles on their own. Therefore, a national stockpile has been created as a resource for all.

The CDC's Strategic National Stockpile (SNS) is a national repository of antibiotics, chemical antidotes, antitoxins, life-support medications, IV administration, airway maintenance supplies, and medical/surgical items to protect the American public if there is a public health emergency (terrorist attack, flu outbreak, earthquake) severe enough to cause local supplies to run out. Once Federal and local authorities agree that the SNS is needed, the SNS will supplement and re-supply state and local public health agencies anywhere and at anytime within the USA or its territories.

The SNS is organized for flexible response. The first line of support lies within the immediate response 12-h Push Packages. These are caches of pharmaceuticals, antidotes, and medical supplies designed to provide rapid delivery of a broad spectrum of assets for an ill defined threat in the early hours of an event. These Push Packages are positioned in strategically located, secure warehouses ready for immediate deployment to a designated site within 12 h of the federal decision to deploy SNS assets. These 12-h Push Packages have been configured to be immediately loaded onto either trucks or commercial cargo aircraft for the most rapid transportation. Concurrent to SNS transport, the SNS Program will deploy its Stockpile Service Advance Group (SSAG). The SSAG staff will coordinate with state and local officials so that the SNS assets can be efficiently received and distributed upon arrival at the site.

If the incident requires additional pharmaceuticals and/or medical supplies, follow-on vendor managed inventory (VMI) supplies will be shipped to arrive within 24–36 h. If the agent is well defined, VMI can be tailored to provide pharmaceuticals, supplies and/or products specific to the suspected or confirmed agent(s). In this case, the VMI could act as the first option for immediate response from the SNS Program.

The SNS Program works with governmental and nongovernmental partners to upgrade the nation's public health capacity to respond to a national emergency. Critical to the success of this initiative is ensuring capacity is developed at federal, state, and local levels to receive, stage, and dispense SNS assets.

In our work on modeling and optimization of public health infrastructure and mass dispensing strategies (Lee et al. 2009a, b), we describe the importance of collaboration among federal, state, local, and private sectors for successful response to large-scale emergency scenarios, and report public/private solutions for meeting the Strategic National Stockpile dispensing requirements as prescribed within the Cities Readiness Initiative program.

8.3.5 Communication and Public Information

8.3.5.1 Communication and Information Technology

Communication programs and infrastructure for emergency response are critical for successful operations (Lee et al. 2009c, 2011b). Communication efforts should be coordinated, planned, and tested. Regular training on usage should be performed.

Briefly, multi-faceted communication infrastructures are needed for effective medical and emergency response. Two-way communication lines must be kept open between emergency/medical responders and commanders for coordination and facilitation of the response effort; broadcast communication lines must be available to inform the general public about medical precautions and available services; and phone service and call answering stations must be maintained to allow the public to report critical emergency situations. Further, communication infrastructure of hospitals and related medical care entities are critical for any emergency response.

Developing and maintaining a robust communication infrastructure among hospitals, emergency medical services and other health-care facilities, as well as private medical transport companies and medical supplies vendors, is vital to providing relief operations and services during emergency situations. Such infrastructure provides the core foundation for basic knowledge sharing such as patient volume and severity, emergency room capacity, hospital bed availability; special wards availability; medical personnel specialties; transport vehicle availability; and blood, medicine, and medical supplies inventories. During an emergency situation, such critical information can be communicated to a regional coordinating control center which can assess available resources, identify surpluses and shortages, and coordinate distribution efforts. In fact, building capacity for an interoperable communication system for emergency response is identified as one of the areas that must be prioritized according to The Hospital Preparedness Program established by the US Department of Health and Human Services.

Previous instances of emergency situations such as the September 11, 2001, terrorist attack and the Rhode Island night club fire incident have brought forth the challenges and necessity in establishing and sustaining a redundant emergency communication network. Kapucu 2006 has examined the importance and challenges in establishing a coordinated inter-agency communication network and the role information technology serves to enhance such communication and decision-making as in the context of the September 11, 2001, terrorist attack. Challenges include interoperability among the various communication tools, reliability of the tools during emergency situations, technical limitations, barriers in inter-agency communication and related aspects.

8.3.5.2 Public Information and Risk Communication

Risk communication plays a crucial role to any successful emergency response. The social challenges that arise as a result of human behavioral patterns cannot be over-emphasized. During the high concern and high stress situation, the messages to the public must be concise, unified, and coordinated. Getting correct and credible information out quickly assures the public that they can trust the authority in protecting themselves and their families. The public needs to know in timely manner that there is a problem and the nature of the problem. Appropriate measures must be performed to curtail rumors that may cause unnecessary panic and fear, and potential unrest. Multi-media hotlines should be established to share factual information and timely updates. Educational information pertinent to the medical countermeasures and dispensing sites should be clearly stated. Finally, care must be taken to ensure that vulnerable and special need populations are served appropriately.

8.3.6 Lessons Learnt and Continued Challenges

8.3.6.1 Our Experience Through Anthrax Drills and Actual Vaccination Events

Our experience illustrates that by combining mathematical modeling, large-scale simulation, and powerful optimization engines, and coupling these with automatic graph-drawing tools and a user-friendly interface, we can design and implement a fast and practical emergency response decision support tool that can run on a wide range of computing platforms, including PDAs for real-time usage. The system, developed in collaboration with CDC SNS investigators, offers public health emergency coordinators the capabilities to

1. Determine strategically most effective locations for POD facilities to best serve the affected population.
2. Design customized and efficient POD floor plans via an automatic graph-drawing tool. Users can design and compare various floor plans to determine the tradeoffs in personnel usage as well as operations efficiency.
3. Determine optimal labor resources required and provide the most-efficient placement of staff at individual stations within the POD. The resulting staffing plans maximize the number of individuals who can be treated, minimize the average time patients spend in the clinic, and equalize utilization across clinic stations.
4. Perform disease propagation analysis, understand and monitor the intra-POD disease dilemma, and help to derive dynamic response strategies to mitigate casualties. The what-if analysis and worst-case scenarios can also equip epidemiologists with knowledge that may assist emergency planners in testing alternative POD facility layouts, assess batch size for patient orientation, and analyze the tradeoffs between POD throughput operations and degree of infectiousness.
5. Assess current resources and determine minimum needs to prepare for readiness in emergency situations for their regional population.
6. Carry out large-scale virtual drills and performance analyses, and investigate alternative strategies.
7. Train personnel, and design emergency exercises with a variety of dispensing scenarios. Such training exercises could be used to quickly get new emergency preparedness planners up to speed and to keep existing planners sharp.

The computational advances also provide flexibility to quickly analyze design strategies and decisions, and can generate a feasible regional dispensing plan (with a network of cost-effective PODs, each operating at various throughput rates and utilizing various dispensing modalities) based on the best estimates and analyses available, and then allow for reconfiguration of various PODs as the event unfolds.

Some of the regional planning and operational analysis reveal that

1. Sharing labor resources across counties and districts within the same jurisdiction is important.
2. The most cost-effective dispensing plan across a region consists of a combination of drive-through, walk-through, and closed PODs, each operating at a throughput rate that depends on the surrounding population density, facility type, and labor availability.
3. The optimal combination of POD modalities changes according to various facility capacity restrictions, and the availability of critical public health personnel.
4. An increase in the number of PODs in operation does not necessarily increase the total number of core public health personnel needed.

5. Optimal staffing is nonlinear with respect to throughput; thus we cannot estimate the optimal staffing and throughput by simply using an average estimate.
6. Depending on the population, an "optimal" capacity that provides the most effective staffing exists for each POD location. If a POD is operating above its optimal capacity, reduction in capacity (and thus hourly throughput) eases the crowd-control tasks of law-enforcement personnel and helps to minimize potential operational problems inside the POD.

RealOpt has been used successfully by over 4,000 public health and emergency directors and coordinators in planning for biodefense drills (e.g., anthrax, smallpox) and pandemic response events in various locations in the USA since 2005. It has also been used for dispensing clinic design and staff allocation for the 2009 H1N1 mass vaccination campaign. Users have tested various POD layouts, including drive-through, walk-through, and closed PODs. Because of the system's rapid speed, it facilitates analysis of "what-if" scenarios, and serves not only as a decision tool for strategic and operational planning, actual drill preparation, and personnel training, but also allows dynamic reconfigurations as an emergency event unfolds. In addition, it supports performing "virtual field exercises," offering insight into operation flows and bottlenecks when mass dispensing is required.

8.3.6.2 Continued Challenges and the Need for Multi-layer Protection

The ability to analyze planning strategies, compare the various options, and determine the most cost-effective combination dispensing strategy are critical to the ultimate success of mass dispensing.

The federal government continues to seek advances in this challenging area. In the 2009 Executive Order, signed on December 30, 2009 by the President of the USA, a policy was setup to plan and prepare for the timely provision of medical countermeasures to the American people in the event of a biological attack in the USA through a rapid federal response in coordination with state, local, territorial, and tribal governments. The policy seeks to (1) mitigate illness and prevent death; (2) sustain critical infrastructure; and (3) complement and supplement state, local, territorial, and tribal government medical countermeasure distribution capacity.

Specifically, the executive order stipulates that the federal government shall pursue the establishment of a national US Postal Service medical countermeasures dispensing model to respond to a large-scale biological attack with anthrax as the primary threat consideration. Further the federal government must develop the capacity to anticipate and immediately supplement the capabilities of affected jurisdictions to rapidly distribute medical countermeasures following a biological attack by establishment of a federal rapid response capability. The executive order also asks for Continuity of Operations where the federal government must establish mechanisms for the provision of medical countermeasures to personnel performing mission-essential functions to ensure that mission-essential functions of federal agencies continue to be performed following a biological attack.

Postal delivery of medical countermeasures has been discussed in detail in Wein 2008. Because postal workers deliver mail service to all households on a regular basis, the possibility of their usage for the first 12-h initial dispensing is appealing. This will allow time for the establishment of PODs for continued dispensing service. The availability of personnel during crisis can be volatile, as articulated in our medical surge discussion (Lee et al. 2011a). However, such a multi-layer dispensing plan allows flexibility in response as well as leveraging and integration of heterogeneous resources for maximum coverage and outcome. Postal and POD distribution strategies require different levels of security; their tradeoffs and complimentary characteristics should be carefully analyzed.

8.4 Large-Scale Disaster Relief Efforts

Large-scale disaster (humanitarian) relief efforts (e.g., in response to earthquakes, hurricanes, forest fires) where homes are destroyed, critical infrastructures are damaged, and tens of thousands or millions of people's lives are affected, require rapid establishment of "*service constructs*." These service constructs serve as home shelters for the population being displaced; as distribution nodes for receiving supplies for on-the-ground responders; as dispensing sites for handing out food and water to the affected population; and as hospital tents for medical care of the sick. In the aftermath of such an event, the medical surge requirement is acute (Lee et al. 2011a), evacuation orders are highly probable (Lee et al. 2011c), and the need for effective communication and coordination for humanitarian relief effort is crucial (Lee et al. 2009c, 2011b).

We highlight below various issues that are prominent within disaster relief efforts.

8.4.1 Network of Service Constructs, Staffing and Resource Constraints, Opportunistic Disease Spread, and Supply Chain Management

The scope and complexity of establishing service constructs (for food and water, shelters, and medical care, etc.) are very similar to those for large-scale mass dispensing efforts. Past disaster relief efforts such as in the aftermath of the 2010 earthquake in Haiti highlight the daunting challenges as the regional response must rapidly establish (in an ad hoc manner) dispensing and distribution networks. Many elements relevant to mass dispensing and population protection come into play. Further, the lack of water, electricity, shelters, poor sanitation, and the existence of at most a barebones critical infrastructure present enormous challenges.

Staffing and resources are often under severe shortage where ad hoc skills and just-in-time training are provided. Skilled workers such as those with medical background, logistics and operational, and security experiences may often be deployed from other countries and nonprofit organizations to help with rapid response, relief, and rebuild mission.

Crowded, unsanitary conditions in improvised refugee shelters could spread illnesses such as typhoid and measles. Thus, mass prophylaxis treatments are often needed to prevent disease spread, even though illness and infections can still be a threat to survivors. Puddles of filthy water accumulated near service constructs may become a breeding ground for mosquitoes. That, in turn, may lead to the spread of deadly diseases and epidemics such as dengue, malaria, and cholera. Disease mitigation and facility layout strategies are absolute critical tasks to the livelihood of these survivors.

In disasters, suppliers and stakeholders can be very diverse, and there is usually no unifying business and operating theme to manage them. Further, unsolicited and unwanted donations can burden the management team, causing overload of arrivals at sea or airports, or congesting the warehouses (Chomilier et al. 2003; Cassidy 2003; Murray 2005).

Routing available and necessary goods from entry ports (sea or air) to affected sites can pose daunting challenges. Efficient usage of potentially limited sea/air space and landing sites, available roads, vehicles, fuels, drivers, and material handling equipment at the receiving ends are all uncertain and unreliable. Food safety, sanitation hygiene, and opportunistic looting further complicate the process.

Effective coordination of multiple humanitarian agencies, in addition to military, government, and private entities, is of great importance in response to disasters. This is both a challenge and an opportunity given the differing missions, histories, and expertise of these institutions.

8.4.2 Coordination Among Different Stakeholders

Humanitarian operations often have a large number and variety of stakeholders and multiple organizations operating in the same place simultaneously but without (formal) coordination. A loosely coupled coordination of different aid agencies, suppliers, and local and regional actors, each has their own way of operating and own organizational structures, can work charmingly, yet it can also pose discord (Long and Wood 1995). The different political agendas, ideologies and religious beliefs, and the need for appeal to public for donations and media attention complicate the work. Perhaps, the greatest challenge here lies in aligning these organizations properly without compromising their mandates and beliefs. Further, people from different cultural background may have different traditions that may hinder the communication and coordination between organizations (Van Wassenhove 2006). As seen in the recent Haiti event, operational and organizational differences can also create frictions and misunderstanding among various responding nations, thus masking the effectiveness of the operations.

Sometimes, the affected areas may not be reachable due to political reasons. For example, after the 2005 South Asian Tsunami, the Indonesian government felt compelled to allow free entry in a region that had been very restricted for a long time. This caused a huge influx of personnel from humanitarian organizations, ad hoc organizations, and volunteers on sites, overcrowding the sites.

8.4.3 Risk and Uncertainty

Personnel working within a disaster response environment are often exposed to destabilized infrastructure (Cassidy 2003; Murray 2005). Not only do they need to work in facilities or areas that are physically damaged, the social effect taking place after a disaster could often become overwhelming. Many disasters such as tsunamis and earthquakes have after-effects that could further cause panic or disruption to response operations. Uncertainty in risk, in when the suppliers will arrive and where, in the amount of supplies that are available, in the overwhelming demand needed by the affected population, add layers of anxiety and stress to these on-the-ground workers. "Disaster relief supply chain" shows the extremes of a trend toward more uncertainty and risk prevalent in today's global business supply chain.

8.4.4 Media Exposure

Disaster events often have high exposure to the media. However their relationship is described by some as a love–hate one (Van Wassenhove 2006). On the positive side, high exposure to the media means more public attention on the affected areas. This often translates into more donations and general support. However, mass media is short-sighted and tends to be more interested in catastrophic events while putting less focus on long-term humanitarian and relief effort. There is a need to educate and collaborate with news media personnel on disseminating long-term challenges and efforts to the general public. In recent years, some improvement has been seen as news media and public have been made aware of response failures in the emergency responses related to Katrina and other natural disasters, and the high-profile public scrutiny of the federal and local response effort. In general, appropriate media usage can educate the public and help spread word of the situation so as to garner financial donations that are much needed in any disaster response scenario.

8.5 Summary

8.5.1 *Optimization, Simulation, and Dynamical Systems Methodologies*

Operations research, with its roots in defense and military operations, has a natural place in emergency response planning and execution. Optimization, stochastic, simulation, systems modeling, and decision analysis approaches are routinely used to plan for and aid in analyzing a broad spectrum of emergency responses as a result of natural or man-made disaster (Larson 1975; Larson et al. 2006; Green and Kolesar 2004; Lee et al. 2006a, 2009b). Specifically, Green and Kolesar (2004) trace the history of operations research and management science applications in emergency response, with a particular focus on the work done in New York City between 1969 and 1989. Many projects were undertaken by a group of researchers as part of the New York City-RAND Institute (NYCRI) initiative. These included applications to ambulance, fire, and police car location and deployment.

Resource allocation, scheduling, facility location, vehicle routing, inventory control, and transportation logistics in emergency response have all been formulated into optimization models. Among the resource allocation models, Fiedrich et al. (2000) uses dynamic optimization model for the initial search-and-rescue period after a strong earthquake. Branas et al. (2000) used a trauma resource allocation model for ambulances and hospitals. Tzeng et al. (2007) present a multi-objective optimal planning model for designing relief delivery systems. The three goals are minimizing the total cost, minimizing the total travel time, and maximizing the minimal satisfaction during the planning period. Yan and Shih (2007) use a time-space network model (based on an integer network flow problem with side constraints) to minimize the length of time needed for emergency repair, with related operating constraints for emergency repair work team scheduling.

Lee et al. (2006a, 2009a) investigate resource allocation, staffing, facility location, and multi-modality mass dispensing strategies and emergency response for biodefense and infectious disease outbreaks. Some of the integer programming instances include on the order of ten million variables, and the authors provide rapid solution engines to arrive within 5% to optimality under 3 min. Zhang and Yang (2007) present an optimization model and algorithm of a facility location problem in a perishable commodities emergency system. Doerner et al. (2009) present a model for multi-objective decision analysis with respect to the location of public facilities such as schools in areas near coasts, taking risk of inundation by tsunamis into account. Coskun and Erol (2010) use an integer optimization model to decide locations and types of service stations, and regions covered by these stations under service constraints in order to minimize the total cost of the overall system. The model can produce optimal solutions within a reasonable time for large cities having up to 130 districts or regions.

Many authors present routing and transportation studies. Harewood (2002) uses a multi-objective version of the maximum availability location problem to determine emergency ambulance deployment. Wang et al. (2009) analyze and propose the concept of post-earthquake road safety, dividing emergency vehicle routing choice and optimization problem into two decision-making stages: pre-trip and en-route. Sheu (2007) uses a hybrid fuzzy clustering-optimization approach to study the operation of emergency logistics co-distribution when responding to urgent relief demands in the crucial rescue period. Yi and Kumar (2007) present a meta-heuristic of ant colony optimization for solving a logistics problem arising in disaster relief activities. Yuan and Wang (2009) present two mathematical models for path selection in emergency logistics management. The models include actual factors in time of disaster. Liu and Zhao (2009) model an emergency materials distribution problem in an anti-bioterrorism system as a multiple traveling salesman problem.

Simulation has been used in numerous public health and medical preparedness topics, including evacuation, resource allocation and patient flow, routing in emergency medical services and surge planning, and disease propagation analysis.

Simulation has also been used in resource allocation and patient flow (Hupert et al. 2002, 2003; Lee et al. 2006a, b; Mason and Washington 2003; Wang et al. 2008; Rossetti et al. 1999; Saleh and Othman 2008); and in the area of pandemic response strategies and mitigation (Aaby et al. 2006; Barnes et al. 2009; Das et al. 2007; Hupert et al. 2002; Lee et al. 2006a, b, 2009a, b, c, 2010a; Meltzer et al. 2001; Wang et al. 2008; Wu et al. 2009). It has found many applications in routing in emergency medical services and surge planning (Barnes et al. 2009; Goldberg and Paz 1991; Haghani et al. 2004; Su and Shih 2003).

As modelers attempt to incorporate more realistic dynamics into epidemiology and disease propagation models (such as stochasticity, nonexponential waiting times, sample-path-dependent events, and demographical and geographical data), more flexible tools, such as individual-based stochastic simulations, are preferable. Although simulation is a powerful approach, the resulting models are often mathematically intractable and require advances in computational strategies (Eubank 2002; Gani and Leach 2001; Epstein et al. 2004; Ferguson et al. 2005, 2006; Longini et al. 2005; Germann et al. 2006; Lee et al. 2009b, 2010a).

The discipline of dynamical systems, mostly differential equation systems, provides the principal methods of modeling in classical mathematical epidemiology (Anderson et al. 1992; Diekmann and Heesterbeek 2000). It has also been used to analyze strategies and policies. Wein et al. (2003) use a system of differential equations to model the effects of different policies in response to an anthrax bioterror attack. The system includes an atmospheric dispersion model for the spread of the bacterium causing anthrax, an age-dependent dose–response model for the impact of treatment on an individual, a disease progression model to capture the stages through which an infected individual goes, and a set of two-stage queuing systems for antibiotic distribution and hospital care. Kaplan et al. (2002) imbeds the evaluation of vaccination logistics policy within a disease propagation model and compares strategies of traced vaccination, mass vaccination, and the mixed response advocated by the Centers for Disease Control. Eichner et al. (2007) developed a deterministic model for evaluating impact of different intervention strategies during pandemic influenza. The model is based on over 1,000 differential equations which extend the classic SEIR model by clinical and demographic parameters relevant for pandemic preparedness planning. The model aims to operate with an optimal combination of precision, realism, and generality. Wu et al. (2007) develops models to demonstrate that a pre-pandemic vaccine allocation policy that allocates vaccine to each state in proportion to the population size is not the most efficient. In fact, an inequitable strategy that allows no allocation to some regions while the sufficient vaccines being allocated to other regions demonstrates larger benefit. However, if considering other strategy selection criteria such as simplicity, robustness, and equity, the current pro-rata policy is a good compromise.

8.5.2 Integrated Approaches, and Information and Decision Support Systems

The burden of responding to a public health or medical disaster is multi-faceted and a genuine test to the sustainability of critical infrastructure. Negotiating emergency operations is especially difficult due to inter-agency goal conflicts, differences in organizational culture and bureaucratic constraints, discrepancies in situation assessment, scarcity of resources, and organizational complexity. Nevertheless, interplay among many agencies is critical, and consequently, integrated approaches, and information and decision support systems prove to be very beneficial as part of the solution strategies (Kananen et al. 1990; Kwan and Lee 2005; Nguyen et al. 2005; Raghu et al. 2005; Rotz and Hughes 2004; Subramaniam and Kerpedjiev 1998).

Iakovou and Douligeris (2001) present the development of IMASH, an Information Management System for Hurricane disasters. IMASH is an intelligent integrated dynamic information management tool, capable of providing comprehensive data pertaining to emergency planning and response for hurricane disasters. Popp et al. (2004) developed information-analysis tools for an effective multi-agency information-sharing effort.

Zografos et al. (1998) describes an integrated framework consisting of a data management module, a vehicle monitoring and communication module, and a modeling module, for managing emergency response of the electric utility companies. The framework integrates a GIS system with a decision-making modeling module to deliver solutions in real time that optimize deployment of the available emergency resources. El-Anwar et al. (2009) present the development of an automated system to support decision-makers in optimizing post-disaster temporary housing arrangements. The system has been integrated in MAEviz and provides the capability of optimizing a number of important objectives, including minimizing negative socio-economic impacts, maximizing housing safety, minimizing negative environmental impacts, and minimizing public expenditures.

Bui et al. (2000) proposes a framework for developing a global information network (GIN). The application would incorporate four factors that affect the design of a GIN: nature of disaster relief operations, negotiation styles of participants, social, cultural, and organizational characteristics of participants and resource availability. Such a framework could provide a set of basic metrics and factors to characterize any disaster situation. GIN would use high speed internet as backbone, it includes a command center where the disaster management team would be, and telecommunication channels that connect to expert advice groups from around the world. The GIN would also be linked to an array of data and knowledge base warehouses.

NYU's PLAN C is an innovative tool for emergency managers, urban planners and public health officials to prepare and evaluate Pareto-optimal plans to respond to urban catastrophic situations. Doheny and Fraser (1996) describe a software tool for modeling the decisions that people make in emergency situations in offshore environments. It can be used to predict the likely behaviors of a population in hazardous situations and help evaluate the effectiveness of emergency procedures and training.

Mondschein (1994) reviews the use of spatial data by environmental managers and emergency responders who are charged with the responsibility to perform hazard assessments, identify the location of toxic and hazardous materials, deploy emergency resources, and review demographic data to ensure the safety of the public and the surrounding communities.

In our own work, the decision support system RealOpt combines OR modeling techniques, novel and large-scale computational engines, sophisticated graph-drawing tools, 3D geographical spatial information with federal census data, and demographic and socio-economic data for operational and strategic planning, and policy analysis. RealOpt allows public health emergency coordinators to (1) determine locations for service facilities setup; (2) design customized, efficient floor plans for each facility via an automatic graph-drawing tool; (3) determine (in real-time) optimal resource allocation through advanced computational techniques in simulation and optimization; (4) monitor intra-facility disease propagation through a novel disease propagation model, and derive dynamic response strategies to reduce the spread of disease and mitigate the risk of casualties; (5) assess resources and determine minimum needs to prepare for treating regional populations; (6) carry out large-scale virtual drills and performance analyses, and investigate alternative dispensing strategies; and (7) design a variety of dispensing scenarios and emergency-event exercises to train personnel (Lee 2009; Lee et al. 2006a, b, 2009a, b, 2010a).

8.5.3 Challenges

Modeling and optimizing public health infrastructure involve elements of resource allocation under risk, uncertainty, and time pressure; large-scale supply chain management; transportation and operational logistics; and medical treatment and population protection. The operations must be supported by an effective communication infrastructure. There is a necessity for vertical and horizontal integration and communication, where federal, state, local, tribal, territorial, private, and business stakeholders work toward a common goal of a resilient public health system. The infrastructure must be *flexible, scalable, sustainable, and elastic* to support an effective and timely response, and to mount rapid recovery and mitigation operations.

The 2007 Homeland Security Presidential Directive-21 (The White House 2007) establishes a National Strategy for Public Health and Medical Preparedness, which builds upon a four pillar framework—Threat Awareness, Prevention and Protection, Surveillance and Detection, and Response and Recovery. It aims to transform our national approach to protecting the health of the citizens against all disasters. Although the four pillars were developed initially to guide the efforts to defend against a bioterrorist attack, they are applicable to a broad array of natural and man-made public health and medical challenges, and are appropriate to serve as the core functions of the strategy for public health and medical preparedness.

Public health and medical preparedness continue to shower challenges to the scientific community. Some critical issues include (a) realistic systems modeling, (b) intractability of large-scale instances, (c) inter-dependencies among multiple critical components/agencies, and (d) the importance and necessity for end-to-end systems modeling and design. Technological advances are needed to allow for complex realistic modeling while providing users with affordable computational power that result in decision systems that are practical for actual scenario-based analysis. The effective integration and alignment of care personnel, facilities, and equipment and supply for optimal outcome remains essential. Capability to solve large-scale resource allocation and location problems is a must. Tracking of disease and designing and implementing dynamic mitigation strategies will have a tremendous impact on population protection. Supply chain management needs to be dynamic, and multi-agency partnership models should be developed. Information sharing and management, and risk and communication strategies continue to evolve. Multi-modality integration of technologies and reliable platforms for communication and public dissemination are critical. Policy and coordination among different stakeholders across country borders need to be studied and potentially streamlined. While many issues relate to operational and strategic planning, many others involve policies, risk management, security, public communication, and cultural and human behavior.

The key to success is flexibility and adaptability—in staffing, in operations strategies, in coordinating and communications strategies, and in the willingness (for multiple agencies) to collaborate. Initial plans must be put in place and executed rapidly and yet allow reconfiguration on the fly as an event unfolds.

Acknowledgments Part of the material and results reported herein are based on our work in this area and interaction with public health agencies, and discussion with many state and federal public health and emergency response experts. While there are many people to thank in this multi-agency and multi-disciplinary collaboration, the authors would like to specially thank Dr. Jacquelyn Mason of the CDC, and Tom Tubesing, formerly of the CDC, Dr. Lawler and Dr. Mecher, formerly at the Homeland Security Council in the White House, William Glisson at ESi, Bernard Hicks at DeKalb Emergency Preparedness Department, and the many public health and emergency managers throughout the nation. We also acknowledge the funding from the CDC and Defense Threat Reduction Agency.

References

Aaby K, Herrmann JW, Jordan CS, Treadwell M, Wood K (2006) Montgomery County's public health service uses operations research to plan emergency mass dispensing and vaccination clinics. Interfaces 36(6):569–579

Aly AA, White JA (1978) Probabilistic formulation of the emergency service location problem. J Oper Res Soc 29(12):1167–1179

Anderson RM, May RM, Anderson E (1992) Infectious disease of human: dynamics and control. Oxford University Press, Oxford

Barnes AJ, Jacobson JO, Solomon MD, Kun H. Eugene Grigsby JI (2009) Los Angeles County pandemic flu hospital surge planning model. National Health Foundation, Los Angeles, California

Berman OA, Gavious A (2007) Location of terror response facilities: a game between state and terrorist. Eur J Oper Res 177(2):1113–1133

Branas CC, MacKenzie EJ, ReVelle CS (2000) A trauma resource allocation model for ambulances and hospitals. (Managerial and Policy Impact). Health Serv Res 35(2):489–507

Brandeau ML, Chiu SS (1989) An overview of representative problems in location research. Manag Sci 35(6):645–674

Bui T, Cho S, Sankaran S, Sovereign M (2000) A framework for designing a global information network for multinational humanitarian assistance/disaster relief. Inform Syst Front 1:427–442

Cassidy W (2003) A logistics lifeline. Traffic world, 267(43):15

Chaiken JM, Larson RC (1972) Methods for allocation urban emergency units: a survey. Manag Sci 19(4):P110–P130

Chomilier B, Samii R, Wassenhove LNV (2003) The central role of supply chain management at IFRC. Forced Migration Rev 18:15–18

Church RL, Scaparra MP, Middleton RS (2004) Identifying critical infrastructure: the median and covering facility interdiction problems. Ann Assoc Am Geogr 94(3):491–502

Coskun N, Erol R (2010) An optimization model for locating and sizing emergency medical service stations. J Med Syst 34:43–49

Das TK, Savachkin AA, Zhu Y (2007) A large scale simulation model of pandemic influenza outbreaks for development of dynamic mitigation strategies. IIE Trans 40(9):893–905

Diekmann O, Heesterbeek J (2000) Mathematical epidemiology of infectious diseases: model building. Wiley, New York

Doerner KF, Gutjahr WJ, Nolz PC (2009) Multi-criteria location planning for public facilities in tsunami-prone coastal areas. OR Spectr 31:651–678

Doheny JG, Fraser JL (1996) MOBEDIC: a decision modelling tool for emergency situations. Expert Systems Appl 10:17–27

Eichner M, Schwehm M, Duerr H-P, Brockmann SO (2007) The influenza pandemic preparedness planning tool InfluSim. BMC Infect Dis 7:17

El-Anwar O, El-Rayes K, Elnashai A (2009) An automated system for optimizing post-disaster temporary housing allocation. Autom Construct 18:983–993

Epstein J, Cummings DAT, Chakravarty S, Singa R, Burke DS, Cummings JD (2004) Toward a containment strategy for smallpox bioterror: an individual-based computational approach. The Brookings InstitutionPress, Washington, DC

Eubank S (2002) Scalable, efficient epidemiological simulation. SAC '02: Proc. 2002 ACM Sympos. Appl. Comput. ACM Press, New York, pp 139–145

Ferguson NM, Cummings DA, Cauchemez S, Fraser C, Riley S et al (2005) Strategies for containing an emerging influenza pandemic in Southeast Asia. Nature 437:209–214

Ferguson NM, Cummings DAT, Fraser C, Cajka JC, Cooley PC et al (2006) Strategies for mitigating an influenza pandemic. Nature 442:448–452

Fiedrich F, Gehbauer F, Rickers U (2000) Optimized resource allocation for emergency response after earthquake disasters. Saf Sci 35:41–57

Gani R, Leach S (2001) Transmission potential of smallpox in contemporary populations. Nature 414(6865):748–751

Gerberding JL (2003) Testimony before the United States Senate Committee on Health, Education, Labor, and Pensions, January 30, 2003. http://www.hhs.gov/asl/testify/t030130.html. Accessed 3 October 2012

Germann TC, Kadau K, Longini IM, Macken CA (2006) Mitigation strategies for pandemic influenza in the United States. Proc Natl Acad Sci U S A 103:5935–5940

Goldberg J, Paz L (1991) Locating emergency vehicle bases when service time depends on call location. Trans Sci 25(4):264–280

Green LV, Kolesar PJ (2004) ANNIVERSARY ARTICLE: improving emergency responsiveness with management science. Manag Sci 50:1001–1014

Haghani A, Tian Q, Hu H (2004) Simulation model for real-time emergency vehicle dispatching and routing. Trans Res Rec 1882:176–183

Harewood S (2002) Emergency ambulance deployment in Barbados: a multi-objective approach. J Oper Res Soc 53:185–192

Hogan K, Revelle C (1986) Concepts and applications of backup coverage. Manag Sci 32 (11):1434–1444

Hupert N, Mushlin AJ, Callahan MA (2002) Modeling the public health response to bioterrorism: using discrete event simulation to design antibiotic distribution centers. Med Decis Mak 22 (5 suppl):S17–S25

Hupert N, Bearman GML, Mushlin AI, Callahan MA (2003) Accuracy of screening for inhalational anthrax after a bioterrorist attack. Ann Int Med 139(5):337–345

Iakovou E, Douligeris C (2001) An information management system for the emergency management of hurricane disasters. Int J Risk Assess Manag 2:243–262

Jia H, Ordonez F, Dessouky M (2002) A modeling framework for facility location of medical services for large-scale emergencies. IIE Trans 39(1):41–55

Kananen I, Korhonen P, Wallenius J, Wallenius H (1990) Multiple objective analysis of input–output models for emergency management. Oper Res 38:193–201

Kaplan EH, Craft DL, Wein LM (2002) Emergency response to a smallpox attack: the case for mass vaccination. Proc Natl Acad Sci U S A 99:10935–10940

Kapucu N (2006) Interagency communication networks during emergencies: boundary spanners in multiagency coordination. Am Rev Public Adm 36:207–225

Kermack WO, McKendrick AG (1991) Contributions to the mathematical theory of epidemics—III. Further studies of the problem of endemicity. Bull Math Biol 53:89–118

Kwan M-P, Lee J (2005) Emergency response after 9/11: the potential of real-time 3D GIS for quick emergency response in micro-spatial environments computers. Environ Urban Syst 29:93–113

Larson RC (1975) Approximating the performance of urban emergency service systems. Oper Res 23(5):845–868

Larson RC, Metzger MD, Cahn MF (2006) Responding to emergencies: lessons learned and the need for analysis. Interfaces 36(6):486–501

Lawler JV, Mecher CE (2007) Homeland Security Council, private communication

Lee EK (2008) Doing good with good O.R O.R.s Do-Gooders. National biodefense–in case of emergency. OR/MS Today 35(1):28–34

Lee (2009) A systems view of POD operations: integrating all the elements. Presentation at the Institute of Medicine Workshop on Medical Countermeasures Dispensing: Emergency Use Authorization. Washington, D.C., November 18

Lee EK, Maheshwary S, Mason J, Glisson W (2006a) Large-scale dispensing for emergency response to bioterrorism and infectious disease outbreak. Interfaces 36(6):591–607

Lee EK, Maheshwary S, Mason J, Glisson W (2006b) Decision support system for mass dispensing of medications for infectious disease outbreaks and bioterrorist attacks. Ann Oper Res Comput Optim Med Life Sci 148:25–53

Lee EK, Chen CH, Pietz F, Benecke B (2009a) Modeling and optimizing the public health infrastructure for emergency response. Interfaces 39(5):476–490

Lee EK, Smalley HK, Zhang Y, Pietz F, Benecke B (2009b) Facility location and multi-modality mass dispensing strategies and emergency response for biodefense and infectious disease outbreaks. Int J Risk Assess Manag 12(2/3/4):311–351

Lee EK, Yang AY, Chinnappan SG, Guilford TW (2009c) Survey and analysis of emergency communication infrastructure among Georgia State Hospitals. Georgia Division of Public Health, Atlanta, Georgia

Lee EK, Chen CH, Pietz F, Benecke B (2010a) Disease propagation analysis and mitigation strategies for effective mass dispensing. Proc Am Med Inf Assoc 2010:427–431

Lee EK, Yuan F, Pietz F, Benecke B (2010b) Strategies for vaccine prioritization. Presentation at Modeling for Public Health Action: From Epidemiology to Operations Modeling for Public Health, Atlanta, Georgia. December 9–10

Lee EK, Yang Y, Pietz F, Benecke B (2011a) Public health, emergency response, and medical preparedness l: medical surge. In: Cochran JJ, Cox LA, Keskinocak P, Kharoufeh JP, Smith JC (eds). Wiley Encyclopedia of Operations Research and Management Science. Wiley, Hoboken, New Jersey

Lee EK, Guilford T, Yang Y, Pietz F, Benecke B (2011b) Public health, emergency response, and medical preparedness III: communication infrastructure. In: Cochran JJ, Cox LA, Keskinocak P, Kharoufeh JP, Smith JC (eds). Wiley Encyclopedia of Operations Research and Management Science. Wiley, Hoboken, New Jersey

Lee EK, Yang Y, Pietz F, Benecke B (2011c) Public health, emergency response, and medical preparedness IV: emergency evacuation. In: Cochran JJ, Cox LA, Keskinocak P, Kharoufeh JP, Smith JC (eds). Wiley Encyclopedia of Operations Research and Management Science. Wiley, Hoboken, New Jersey

Lee EK, Pietz F, Benecke B, Mason M, Burel G (2013) Advancing public health and medical preparedness with operations research. 2012 Franz Edelman Award Issue. To appear 2013

Liu M, Zhao L (2009) Optimization of the emergency materials distribution network with time windows in anti-bioterrorism system. Int J Innov Comput Inf Control 5:1349–4198

Long DC, Wood DF (1995) The logistics of famine relief. J Bus Logist 16:213–229

Longini IM Jr, Nizam A, Xu S, Ungchusak K, Hanshaoworakul W, Cummings DA, Halloran ME (2005) containing pandemic influenza at the source. Science 309:1083–1087

Mason J, Washington M (2003) Optimizing staff allocation in large-scale dispensing centers. CDC report

Meltzer M, Damon I, LeDunc JW, Miller DJ (2001) Modeling potential responses to smallpox as a bioterrorist weapon. Emerg Infect Dis 7(6):959–969

Mondschein LG (1994) The role of spatial information systems in environmental emergency management. J Am Soc Inf Sci 45:678–685

Murray S (2005) How to deliver on the promises. Financial Times. January 6

Narasimhan S, Pirkul H, Schilling DA (1992) Capacitated emergency facility siting with multiple levels of backup. Ann Oper Res 40(1):323–337

Nguyen S, Rosen J, Koop C (2005) Emerging technologies for bioweapons defense. Stud Health Tech Informat 111:356–361

Owen SH, And Daskin MS (1998) Strategic facility location: a review. Eur J Oper Res 111 (3):423–447

Pirkul H, Schilling DA (1988) The siting of emergency service facilities with workload capacities and backup service. Manag Sci 34(7):896–908

Popp R, Armour T, Senator T, Numrych K (2004) Countering terrorism through information technology. Commun ACM 47:36–43

Raghu TS, Ramesh R, Whinston AB (2005) Addressing the homeland security problem: a collaborative decision-making framework. J Am Soc Inform Sci Tech 56(3):310–324

Rossetti MD, Trzcinski GF, Syverud SAPA, Farrington HB, Nembhard DTS, Evans GW (ed) (1999) Emergency Department Simulation and determination of optimal attending physician staffing schedules. In: Farrington PA, Nembhard HB, Sturrock DT, Evans GW (eds) Proceedings of the 1999 Winter Simulation Conference, Phoenix, Arizona. December 5–8

Rotz LD, Hughes JM (2004) Advances in detecting and responding to threats from bioterrorism and emerging infectious disease. Nat Med 10:S130–S136

Saleh MS, Othman ZA, Zin AM (2008) ASRTS: An agent-based simulator for real-time schedulers. In: Al-Dabass D, Turner S, Tan G, Abraham A (eds) Proceedings of the Second Asia International Conference on Modelling & Simulation, Kuala Lumpur, Malaysia, May 13–15

Sheu J-B (2007) An emergency logistics distribution approach for quick response to urgent relief demand in disasters. Transport Res E Logist Transport Rev 43:687–709

Su S, Shih C-L (2003) Modeling an emergency medical services system using computer simulation. Int J Med Inform 72:57–72

Subramaniam C, Kerpedjiev S (1998) Dissemination of weather information to emergency managers: a decision support tool. IEEE Trans Eng Manag 45:106–114

Swain R, ReVelle C, Toregas C, Bergman L (1971) The location of emergency service facilities. Oper Res 19(6):1363–1373

The Buffalo News (2008) Drive-through vaccination effort a success in Amherst: 1,385 people receive hepatitis A booster, 22 Sep

The White House (2007) Homeland security presidential directive 21 [HSPD-21]: public health and medical preparedness. http://www.fas.org/irp/offdocs/nspd/hspd-21.htm. Accessed 27 Feb 2010

Tzeng G-H, Cheng H-J, Huang TD (2007) Multi-objective optimal planning for designing relief delivery systems. Transport Res E Logist Transport Rev 43:673–686

Van Wassenhove L (2006) Blackett memorial lecture humanitarian aid logistics: supply chain management in high gear. J Oper Res Soc 57:475–489

Wang J, Yu H, Luo J, Sui J (2008) Medical treatment capability analysis using queuing theory in a biochemical terrorist attack. In: Zhang XS, Liu DG, Wang Y (eds) Proceedings of the 7th International Symposium on Operations Research and Its Applications (ISORA 2008), Lijiang, China, October 31–November 3

Wang J, Hu X, Xie B (2009) Emergency vehicle routing problem in post-earthquake city road network. In: Peng Q, Wang KCP, Qiu Y, Pu Y, Luo X, Shuai B (eds) Proceedings of the International Conference on Transportation Engineering 2009, Chengdu, China, July 25–27

Wein LM (2008) Neither snow, nor rain, nor anthrax. The New York Times, October 13, 2008. http://www.nytimes.com/2008/10/13/opinion/13wein.html?_r=1. Accessed 22 Feb 2010

Wein LM, Craft DL, Kaplan EH (2003) Emergency response to an anthrax attack. Proc Natl Acad Sci U S A 100:4346–4351

Wu JT, Riley S, Leung GM (2007) Spatial considerations for the allocation of pre-pandemic influenza vaccination in the United States. Proc R Soc Lond B Biol Sci 274:2811–2817

Wu JT, Leung GM, Lipsitch M, Cooper BS, Riley S (2009) Hedging against antiviral resistance during the next influenza pandemic using small stockpiles of an alternative chemotherapy. PLoS Med 6:1–11

Yan S, Shih Y-L (2007) A time-space network model for work team scheduling after a major disaster. J Chin Inst Eng 30:63–75

Yi W, Kumar A (2007) Ant colony optimization for disaster relief operations. Transport Res E Logist Transport Rev 43:660–672

Yuan Y, Wang D (2009) Path selection model and algorithm for emergency logistics management. Comput Ind Eng 56:1081–1094

Zhang M, Yang J (2007) Optimization modeling and algorithm of facility location problem in perishable commodities emergency system. In: Lu H, Qiu N, Tang Y (eds) Proceedings of the Third International Conference on Natural Computation (ICNC 2007), Haikou, Hainan, China, August 24–27

Zografos KG, Douligeris C, Tsoumpas P (1998) An integrated framework for managing emergency-response logistics: the case of the electric utility companies. IEEE Trans Eng Manag 45:115–126

Chapter 9
Disaster Response Planning in the Private Sector and the Role of Operations Research

Özlem Ergun, Gonca Karakus, Paul Kerl, Pinar Keskinocak, Julie L. Swann, Monica Villarreal, and Matthew J. Drake

Abstract Organizations in the private sector such as The Home Depot (THD), Lowe's, Wal-Mart, and Waffle House have become actively involved in the disaster response operations in their communities. With the objective of becoming effective first responders, these companies have integrated a disaster response planning process to their business operations. We introduce the disaster response planning process implemented by THD after their experience with Hurricane Andrew. We describe in detail the components of this process, each of which requires different decisions to be made at different levels of the organization. We discuss how operations research methodologies could be used to assist decision makers in the disaster response setting, and we propose an optimization model for each of two of the decisions commonly found in a disaster response planning process: advance purchasing and inventory allocation. In addition to these exact methods, we suggest a scenario-based approach, which is more intuitive and allows to incorporate objectives that are harder to model, such as the relative value of the supplies.

9.1 Introduction

No one benefits from a closed store after a disaster hits a community. Organizations in the private sector have recognized that and have joined local, regional, national, and international governments and nongovernmental organizations (NGOs) as disaster first responders. These organizations use their expertise in supply chain management as well as their local knowledge to reopen stores and provide access to

Ö. Ergun • G. Karakus • P. Kerl • P. Keskinocak • J.L. Swann • M. Villarreal (✉)
School of Industrial and Systems Engineering, Georgia Institute of Technology, 765 Ferst Drive NW, Atlanta, GA 30332-0205, USA
e-mail: monica.v@gatech.edu

M.J. Drake
Palumbo and Donahue Schools of Business, Duquesne University, Pittsburgh, PA, USA

J.W. Herrmann (ed.), *Handbook of Operations Research for Homeland Security*, International Series in Operations Research & Management Science 183, DOI 10.1007/978-1-4614-5278-2_9,

supplies and services even before Federal Emergency Management Agency (FEMA) begins its response operations (Huffman 2008). It is a win–win situation, since residents of the affected communities might not have access to basic products otherwise, and companies build good relations with their customers and help to restore the market.

In this chapter, we study how the private sector prepares for disaster response. The chapter is organized as follows: the first section provides a short summary on how companies have successfully implemented their disaster response strategies; the next section presents a case study where we discuss in detail the hurricane season planning process of The Home Depot (THD), and where we recognize some of the complexities of a supply chain management under a disaster setting, with its various decisions at different organizational levels; finally, the last section discusses how operations research (OR) methodologies can be used to improve this decision-making process and consequently improve the performance of the disaster response operations. In this last section, we study in detail the advance purchasing and inventory allocation problems. We develop exact solutions through optimization models and describe a game we developed where we tackle these problems through a scenario-based approach. Appendix includes information on how to obtain the case and game files as well as supporting teaching materials.

9.2 Disaster Response Planning in the Private Sector

Waffle House, Wal-Mart, Lowe's, and THD are examples of organizations that have been widely recognized for effectively aligning their business processes to operate when a disaster affects their communities. A company's success in disaster response is measured to a great extent in terms of how long it takes to reach the area, to put people back to work, and to return stores back to operation. This is the case of Waffle House, which has developed, and continuously improved over decades of experience, hurricane response processes where each functional area has clear responsibilities and roles in enabling the stores to a quick recovery in case of a major disaster event (Ergun et al. 2010a). In fact, FEMA has developed an index to assess the situation after a disaster based on the local Waffle House restaurant response. If the local Waffle House is open and serving food from a full menu, it is green. If the Waffle House is open but has a limited menu, it is yellow. If the Waffle House is not open, it is red (King 2011).

After Hurricane Katrina, Wal-Mart was set as an example of what should be done in disaster response operations. They were on site, providing food and supplies even before FEMA was (Barbaro 2005). In order to effectively respond to disasters, Wal-Mart puts its nationwide response center to work, which includes a sophisticated communications system as well as a world-class logistics network. Also after the devastating Hurricane Rita, Wal-Mart even reopened stores in places with no electricity. Requiring its top managers to sit together while coordinating the response operations was crucial for Wal-Mart's success. This level of coordination has been cited as one of the lessons to be learned from the private sector by the government agencies (Hayes 2005).

Home improvement supplies are in high demand before and after a disaster event, such as a hurricane or a tornado, hits a community. People in the affected areas will rely on having access to these products to protect and rebuild their homes and business. This is why the retailers of two biggest home improvement supplies, Lowe's and THD, have both incorporated disaster response planning into their business processes. In the following section we discuss in detail how THD prepares for the hurricane season. We focus on the shortcomings and the lessons learned and the relevance of the coordination among the different functional areas as well as the multiple and often complex decisions that have to be made.

9.3 Case Study: Disaster Response Planning at THD

Founded in Atlanta, Georgia, THD is one of the world's largest home improvement retailers. In 2010, the company ranked as 29th on the Fortune 500 list of US companies, positioning THD as the fourth-largest retailer in the USA and fifth worldwide, with more than 40,000 different SKUs and 1,900 stores. In addition, THD operates 83 warehouses and distribution centers (DCs), 16 import DCs, 9 carton goods facilities, 30 lumber DCs, 10 transit facilities, and 7 global sourcing offices (The Home Depot 2011). Because of its leadership position in the home repair supply market, THD is naturally motivated to be a strong first responder to both natural and man-made disasters. Moreover, hurricane season predominantly affects the southeastern region of the USA, which comprises the home territory of THD. One of the core values of THD is the commitment to being an active contributor to its community; hence, THD strives to provide its customers with products required for home and business repairs following a disaster, such as tarpaulins and construction and cleaning materials.

A home improvement and repair supplier such as THD is able to provide assistance through its established network of local retail stores. However, these stores are usually located within the affected area, so it is highly likely that they will have sustained similar types of damage as the homes and businesses in the region. If THD is going to attain its highest priority of preparing customers for the event and providing adequate supply levels for the post-event cleanup and repair operations, the company must have a detailed operational plan for stocking the stores with inventory and bringing them back on-line as soon as possible after the disaster. In this section we describe THD's operation plan for disaster response. The material in this section was extracted from the Humanitarian Response Planning at THD case study (Ergun et al. 2010b).

9.3.1 The Genesis of a Disaster Response Planning

When Hurricane Andrew devastated the Florida coast in August 1992, THD was a much different company than it is today. The firm operated approximately one-tenth of the stores that it currently has. THD had never experienced a major disaster event in its 13 years of existence, so it was understandable that the company did not have a detailed plan to prepare for and react to an event the magnitude of Hurricane Andrew. An insufficient tactical plan and lack of experience operating under a disaster setting led THD to several operational challenges in the aftermath of Hurricane Andrew:

- The prediction of phases of customer demand during the hurricane season.
- The procurement of products required in the pre-strike and post-strike phases to the affected stores in an accurate and timely manner.
- The establishment of appropriate and ethical prices for pre-strike preparatory items and post-strike repair items.
- The support of adequate workforce levels before and after the event.

THD did not have an accurate forecast of the kind of products that would be in high demand before or after landfall of a hurricane as strong as Andrew or the magnitude of the demand. THD headquarters identified the products and quantities of these products to send to the stores, but much of this hurricane-positioned inventory went unsold because of a poor forecasting ability of THD. The company disposed its unsold inventory through various salvage companies at a significant loss. The logistical challenges that THD faced emphasized the need of an effective tactical plan in place for future disaster events. Management established a policy whereby product demand and damage reports are consistently communicated to THD headquarters. Also, in accordance with its corporate values of "giving back to (its community)" and "doing the right thing," THD established a "no-profit" pricing policy on specific building materials during Hurricane Andrew (Lohr 1992). This no-profit policy consisted of freezing consumer prices on these items even though the wholesale and distribution costs had increased in the short term. THD also limited consumers' purchase quantities for these items to prevent buyers from acquiring large quantities of these items with the purpose of reselling them. THD made this no-profit policy a permanent part of its operational plan during subsequent disaster events (The Home Depot 1992). Along with the product demand and pricing issues, THD faced several human resources challenges commonly present after any severe disaster. Staffing levels are very difficult to maintain after a disaster event when communications are difficult. Therefore, THD set up various teams of personnel to locate all of the store associates after Andrew's landfall. The first priority of these teams was to confirm that the associates and their families were safe. Once this was confirmed, the teams compiled lists of associates that were willing and able to work in the days immediately following the event.

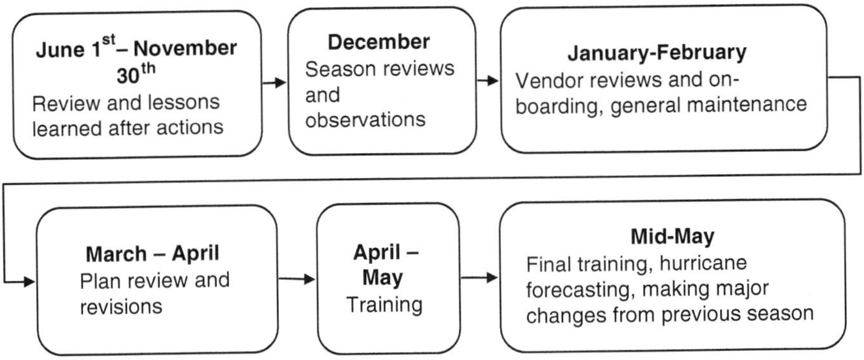

Fig. 9.1 Hurricane season planning schedule (Ergun et al. 2010b)

Hurricane Andrew unfortunately destroyed one of the newest stores of THD in Cutler Ridge, Florida, which had opened 1 week prior to Andrew's landfall. THD set up a circus tent in the parking lot of the Cutler Ridge store and brought in truckloads of products. Several other stores in the area also sustained severe damage, including extensive roof damage and wind/water damage. Much of this damage motivated future efforts of THD on store "hardening" before a hurricane arrives to prepare the store structure.

9.3.2 *Disaster Response Planning Process of THD*

In response to the catastrophic effects of Hurricane Andrew, THD instituted a detailed planning process for the next hurricane season. Over the past years this program has become one of the best-in-class disaster response processes among all types of organizations. In fact, FEMA turned to THD for assistance in re-evaluating its own disaster response process in the wake of the widely publicized difficulties that the organization faced during Hurricane Katrina in September, 2005.

The hurricane season planning cycle, depicted in Fig. 9.1, is centered on the hurricane season in the USA, which is defined to be June 1 through November 30. The planning cycle consists of a thorough discussion and review of the disaster management plans of each of the functional areas of THD, improvement of these plans from year-to-year, and coordination agendas between each of the functional areas for the upcoming hurricane season.

Immediately after the season ends, planning for the next hurricane season begins in December. The functional areas of THD that are largely involved with disaster response operations have post-hurricane season reviews. The first meeting includes each of the key captains of these functional areas: logistics, merchandising, and store operations. In this meeting, actions that worked well and challenges that the company faced during the past season are discussed at a high level. Attendees also identify potential improvements that could be instituted into the plan for the next year. Later in December, individual area meetings for each of the functional areas occur.

In January and February, THD conducts general maintenance and vendor partner reviews. Third-party contractors and merchandise vendors play a crucial role in the success of the disaster response operations of THD. Many of these organizations directly supply stores with products during a disaster situation. Other vendors provide store generators and general store maintenance. THD also works with suppliers of new products to ensure that these vendors can comply with the requirements of the response plan. During March and April, the Manager of Safety Operations/Crisis Management (MSO/CM) examines the revised disaster plans of each of the functional areas. This person ensures the interoperability of these plans and communicates the key linkages and responsibilities among the different areas.

During April and May, training sessions occur, both for the corporate and the field operation teams. The training, which is provided by a third-party organization, includes emergency site assessment, first aid, triage, and emergency management principles. At this time, the emergency call center is supplied with a revised script to use during calls from stores and district managers at the affected areas, during the upcoming hurricane season. Finally, on May 15, the MSO/CM conducts a final meeting with representatives from all the functional areas. This presentation includes a forecast of the events for the upcoming hurricane season and a review of any substantial changes in the disaster response plan from the previous year's plan.

The implementation of the complete hurricane season plan begins as soon as the hurricane season starts on June 1. However, one stage of the plan occurs during the whole hurricane season, when after-action reviews and lessons-learned sessions are held. These sessions can result in a dynamic change of the entire response plan if THD identifies that the current plan is deficient in some way. When changes in the current plan are deemed necessary, THD organizes meetings to communicate any new policies and procedures to the functional areas that are affected.

9.3.3 *Coordination of Disaster Response Operations*

World-class disaster response operations are impossible to implement without a significant commitment of financial, human, and physical capital from the senior management of an organization. THD has made several investments of resources to improve the effectiveness of its response planning process.

9.3.3.1 Storm Forecasting Operations of THD

Accurate predictions of storm frequencies and trajectories are essential for THD to prepare stores to supply their communities with the required products in advance of and after a storm. Before a hurricane season begins, researchers at Colorado State

University produce estimates of seasonal strength. THD uses these high-level forecasts to plan its overall level of relief inventory and transportation service requirements for the coming season. The company also utilizes a third-party weather reporting service called *EarlyAlert* to forecast and track the strength, the trajectory, and projected landfall of every storm. These reports are called "wind field reports." The MSO/CM uses these data to determine when to close stores in advance of a storm and when to open stores after the strike. The wind field reports also form the basis for planning the number of associates that will be required to work in the area during the immediate post-storm period.

9.3.3.2 Disaster Command Center Operations

THD sets up a disaster command center at its corporate headquarters in Atlanta, Georgia, during any Level 3 or higher hurricane situation or in other disaster situations requiring centralized discussion between functional areas. The purpose of the command center is to establish a central coordination point during the disaster event. This command center consists of a central command center, a merchandising and logistics command center, a vendor command center, and a human resources command center. The command centers are equipped with television news feeds from major networks, telephone lines with conference call capabilities, computer workstations with internet access, desks for associates, and other operational requirements. THD also has contingency plans to relocate the command center to an alternate facility in the event of a major disaster at its headquarters.

9.3.3.3 Preparing Regional Facilities for a Disaster Event

As soon as a potential storm is identified, crisis management personnel establish pre-storm communication with stores and other facilities in the affected region through an electronic workload management software application that functions like an e-mail. A package of materials and documents is re-sent to stores in regions most likely to be affected. This package of materials includes pre-strike documents such as IT backup and shutdown procedures, policies for handling claims after the strike, associate financial grant forms, and general operating documents required during the hurricane time frame.

Post-storm conference calls, which are led by the regional vice president (RVP), are held twice daily in storm-affected regional areas. The merchandising team and the RVP review each product category and identify the exact requirements for that particular category during a disaster event. The merchandising representatives then apprise the RVP of any unresolved issues related to product demand and inventory levels.

THD also commits a significant number of resources to care for and support its associates who live in affected areas. In the case they experience any problems or have any questions during a storm, a 24-h emergency line is available for them to call. THD crisis management personnel also handle the post-storm needs of associates in affected areas and schedules blocks of hotel rooms in the event of a disaster for associates that have opted to work at post-strike areas.

9.3.4 Responsibilities of the Functional Areas of THD

Each functional area has a specific role in the disaster response plan of THD. As in the case of its core business, the effectiveness of the relief operations during a disaster event is contingent upon the ability of each functional area to execute its assigned tasks successfully. This section details the core responsibilities of the major functional areas involved in the execution of the disaster response plan.

9.3.4.1 Regional Store Operations

Regional store operations in a disaster situation are executed by a variety of THD personnel. Hourly associates work in stores affected by a disaster, store managers implement the store closing and reopening plans and monitor the procedure checklists, district managers ensure that pre- and post-storm demand information is being communicated accurately to the merchandising personnel, and the RVP makes decisions about closing and opening stores based on local managers' recommendations and wind field reports. As a whole, regional operations perform the following tasks during an emergency event: closing before a storm, opening stores after a storm, preparing stores to withstand storm damage, operating and repairing stores, communicating needs to corporate headquarters, and interacting with local government and community organizations.

9.3.4.2 Distribution and Logistics Operations

When the hurricane season arrives, THD logistics personnel have to ensure business continuity and disaster preparedness of the distribution operations. Items that are in high demand before a storm such as flashlights, batteries, bottled water, and plywood must be delivered effectively from the distribution center to the stores in the storm's path. The logistics department must ensure that these goods are available at the distribution center and can be sent to the stores quickly once it is requested, even when the capacity of the regional transportation network is constrained, for instance, by the local residents' evacuation.

THD utilizes hundreds of different third-party trucking companies to move freight in support of its disaster relief operations and develops special transportation contracts for this purpose. The disaster route contract bidding process at THD occurs separately from that of the normal transportation contracts. These routes are identified as urgent routes, which require carriers to provide a faster delivery and a shorter response time. This bidding process reduces the risk of not finding an available carrier post-storm, and it results in a lower total cost; moreover, this process provides the carriers with better asset planning and utilization.

THD logistics group pre-loads trailers of the products expected to be required both before and after the hurricane strikes, based on the historical usage during similar disaster events. The goods on these pre-loaded trailers are primarily composed of small, low-cost, high-usage, long-shelf-life items such as bottled water instead of large, expensive items such as generators. THD logistics group dispatches these pre-loaded trailers immediately, and then the logistics team works with the RVP and the merchandising team to dynamically plan subsequent loads based on the impact of the storm. Unfortunately, THD distribution centers (DCs) are just as vulnerable to disaster events as the retail stores. For this reason, closing procedures and checklists are also available at each distribution center, and associates at the DCs have the same resources and support services available to retail associates.

9.3.4.3 Merchandising Operations

THD merchandising group is responsible for filling the storage and the transportation space that the logistics group secures with the most critical products for supporting pre- and post-disaster-event operations. Since Hurricane Andrew highlighted the importance of disaster planning, THD has progressively improved its ability to manage the inventory of essential disaster relief products. This expertise, however, has developed slowly over time. After Hurricane Katrina struck New Orleans in 2005, the merchandising group sent large quantities of canned precooked meat, diapers, sunscreen, and mattresses into the affected region. These products turned out to be unnecessary, but on the other hand, the group under-forecasted demand for many other products, and they spent the entire season trying to catch up with the demand. As a result, THD redesigned its merchandising processes within the disaster response plan.

The RVP in the affected region now has control over the particular product mix of the region. This decentralized decision structure assigns product decisions to employees closest to customers, which allows stocking decisions to be made with the most up-to-date, localized demand information. This structure transforms the disaster-relief inventory control system to a pull-based system (as opposed to a push-based system), which allows THD to satisfy demand with a reduced inventory. However, one of the main obstacles to effective disaster response inventory management is the long lead times of some products. THD merchandising group must be able to provide the long-lead-time suppliers with accurate forecasts. A forecast-based order policy hinders THD from using a pull-based strategy with these suppliers.

With the allocation of inventory replenishment decisions to the individual RVPs, the merchandising group now focuses on stocking the four hurricane DCs (located in Puerto Rico, Texas, Florida, and Georgia) and on working with vendors to ensure supply availability. After Hurricane Andrew struck, THD brought supplies into the affected areas on consignment in order to absorb demand shocks as well as to limit demand risk if consumer behavior did not match expectations. The merchandising group now utilizes consignment on 60–70% of the dollar value of its products that support disaster relief operations. Consignment arrangements are especially attractive for products that have a high monetary value, such as chainsaws and generators. Vendors whose products are incorporated into the disaster-relief merchandising plan must be trained in proper packaging methods to support THD operations. These methods include, for instance, shipping products to stores in cartons with lift-off tops to enable store personnel to make the items available for sale immediately, and shipping large items directly on carts to enable customers to wheel them out of the store without employee assistance. The merchandising group has also tried to rationalize the number of SKUs. For example, it is not necessary for THD to provide four or five different brands of AA batteries; it is more important, however, to have enough AA batteries (of any brand) available. Moreover, merchandising personnel work with a smaller number of key vendors instead of ensuring that all of the traditional vendors have a part to play in the plan.

A critical issue for ensuring that internal and external stakeholders focus on the task of serving customers centers on which party should be held responsible for inventory overruns due to disaster relief operations. All of these stakeholders have an interest in helping the affected communities prepare themselves for and rebuild after a disaster event, but none of them want their performance evaluation to be unduly impacted by the merchandising risk that accompanies this operation. The Inventory Planning and Replenishment (IPR) team within the merchandising group establishes a maximum stocking level for each item at each store, and the corporate office will take back any excess supply above this maximum amount. IPR personnel will then redeploy these excess units to other regions in which these items are not currently overstocked.

Availability, timeliness, and accuracy of field information are major determinants of the effectiveness of the THD disaster response plan. Field personnel (e.g., district managers, store managers, and store associates) are in the best position to provide the centralized merchandising group with this up-to-date information because they are working directly in the affected area. THD continuously seeks to improve the accessibility of information from its field operations in order to improve on its already world-class disaster relief operations, for instance, by the use of adequate IT systems.

9.4 The Role of Operations Research in Disaster Planning

OR methodologies have been broadly used to improve supply chain operations. More recently, OR literature on disaster supply chains has grown and caught the attention of both academics and practitioners (The Home Depot 2011). Demand forecasting, inventory planning and allocation, and vehicle routing are just few examples of the problems that need to be analyzed in both traditional and disaster supply chains. However, the highly uncertain setting of a disaster response plus factors such as fairness add complexity to the decision-making process, which makes OR a crucial tool for problem solving in this application area (Altay and Green 2006).

In this section, we present an optimization approach to solve two common problems in disaster response planning faced by first responders such as THD: inventory advance purchasing and allocation.

9.4.1 *The Newsvendor Model for Advance Purchasing Decisions*

Right before and after a disaster strikes, product prices may increase considerably as a result of an increase of demand or a decrease in the supply. Also, transportation infrastructure might be severely damaged, affecting lead times of the delivery of the products. As in THD, a common strategy followed in the public and private sectors is to build inventory in advance. In the case of hurricane season preparedness, public and private organizations use storm forecasts and historical data to predict what and how much is going to be in demand during the coming season. For instance, in the case of hurricanes, the products listed below are in high demand:

- Cover: Blue tarpaulins, plastic sheeting, etc.
- Lighting set: Batteries, spotlight, flashlights, etc.
- Generators
- Bottled water
- Recovery: Rope, ladder, extension cords, rakes, shovels, scoops, etc.
- Cleanup: Cleanup products, sponges, brushes, towels, mops, brooms, buckets, trash cans, trash bags, etc.

Since the pre-storm purchasing decisions occur before the demand is observed, the built inventory can either be insufficient to satisfy the demand or surpass the demand. When deciding on the advance purchasing decisions different trade-offs should be considered. If the inventory is insufficient, there are direct costs related to loss of profit and indirect costs related to loss of customers. On the other hand, products might be purchased at a potential higher cost, which will likely affect profit since policies such as maintaining the customer's regular price are common in companies such as THD. However, if demand is overestimated, inventory will be carried over, and this inventory will incur additional costs, be sold at a lower cost, or be wasted.

We analyze the following setting: there is a single hurricane season which associated demand is uncertain; however, there are historic demand and forecast data that can be used to estimate the probabilistic distribution of the demand. Before the season starts, prices are negotiated with vendors through contracts at discounted prices. After the season starts, products are available at expedited prices. For simplicity, items are grouped in six main categories: cover, lighting, generators, bottled water, recovery, and cleanup. Discounted prices are known for each item group, and they are lower than the expedited prices (also known). Additionally, there is a budget constraint for the pre-season advance orders. This setting can be associated with the Newsvendor problem with a budget constraint.

Newsvendor model is used to decide the optimal inventory level (how much to buy in advance) when the demand during a single time period is uncertain, in order to minimize costs. The intuition behind this model is to balance the overage cost, which is the cost incurred when the demand is less than the inventory, and the underage cost, which is the cost incurred when the demand is greater than the inventory. In this setting, we consider the discounted price as the overage cost because we suppose that the unsold items are wasted (salvage value is zero), and the difference between the expedited and discounted prices as the underage cost because we assume that all unfulfilled demand will be satisfied at the expedited price. Also, we assume that both prices cover the product costs and shipment costs from the vendors to the DCs.

The problem is formulated as a multi-product Newsvendor problem with a budget constraint.

9.4.1.1 Parameters

c_i: Discounted price of item group i, $i = 1, 2, 3, 4, 5, 6$.
p_i: Expedited price of item group i.
$u_i = p_i - c_i$: Unit underage cost of item group i.
$o_i = c_i$: Unit overage cost for item group i.
x_i: Demand of item group i.
$g_i(x_i)$: Probability density function of the demand x_i.
$G_i(x_i)$: Cumulative distribution function of the demand x_i.
B: Budget for advance purchasing.

9.4.1.2 Variables

q_i: Advance purchasing quantity for item group i.

9.4.1.3 Optimization Model

The cost minimization model becomes:

Minimize:

$$\sum_{i=1}^{6}\left[o_i\int_0^{q_i}(q_i - x_i)g_i(x_i)dx_i + u_i\int_{q_i}^{\infty}(x_i - q_i)g_i(x_i)dx_i\right]. \tag{9.1}$$

Subject to:

$$\sum_{i=1}^{6} q_i c_i \le B. \tag{9.2}$$

This model can be transformed into the unconstrained optimization problem of minimizing the following Lagrangian function:

Minimize:

$$L(q_i, \tau) = \sum_{i=1}^{6} \left[o_i \int_{0}^{q_i} (q_i - x_i) g_i(x_i) dx_i + u_i \int_{q_i}^{\infty} (x_i - q_i) g_i(x_i) dx_i \right] + \tau \left[\sum_{i=1}^{6} q_i c_i - B \right]. \tag{9.3}$$

To solve this problem, we compute the partial derivatives of (9.3):

$$\frac{\partial L}{\partial q_i} = \frac{\partial}{\partial q_i} \left[o_i \int_{0}^{q_i} (q_i - x_i) g_i(x_i) dx_i + u_i \int_{q_i}^{\infty} (x_i - q_i) g_i(x_i) dx_i \right] + \tau c_i = 0 \quad \forall i = 1, 2, \ldots, 6 \tag{9.4}$$

$$\frac{\partial L}{\partial \tau} = \sum_{i=1}^{6} q_i c_i - B = 0 \tag{9.5}$$

Using Leibniz rule, (9.4) can be reduced to:

$$\frac{\partial L}{\partial q_i} = o_i \int_{0}^{q_i} g_i(x_i) dx_i + u_i \int_{q_i}^{\infty} g_i(x_i) dx_i + \tau c_i = 0 \quad \forall i = 1, 2, \ldots, 6. \tag{9.6}$$

Next, (9.6) reduces to:

$$\frac{\partial L}{\partial q_i} = o_i G_i(x_i) + u_i(1 - G_i(x_i)) + \tau c_i = 0 \quad \forall i = 1, 2, \ldots, 6. \tag{9.7}$$

Hence, (9.5) and (9.8) should hold for the optimal advance purchasing quantity for each item group q_i^*:

$$G_i(q_i^*) = \frac{u_i - \tau c_i}{o_i + u_i}. \tag{9.8}$$

The algorithm to find optimal quantities is as follows:

- Compute q_i for each item group, given a value of τ. Start with the classic Newsvendor model solution without the budget constraint, where $\tau = 0$.

Table 9.1 Newsvendor model parameters[a]

	Cover set	Lighting set	Generators	Bottled water	Recovery set	Cleanup set
Discounted price, c_i	\$50.00	\$20.00	\$120.00	\$10.00	\$30.00	\$15.00
Expedited price, p_i	\$150.00	\$70.00	\$480.00	\$35.00	\$90.00	\$45.00
Mean (units)	6,000	15,000	3,500	45,000	6,000	7,000
Standard deviation (units)	1,500	4,800	1,000	13,000	1,800	2,175

[a]The data shown do not belong to THD nor any company or organization. Data are entirely hypothetical and were defined only for illustrative purposes

- If the obtained quantities q_i do not violate the budget constrain, i.e., $\sum_{i=1}^{6} q_i c_i \leq B$, then the solution is feasible and optimal for the original constrained problem.
- If the budget constraint is violated, then the optimal solution is found by iterating on the value of τ:
  - If $\sum_{i=1}^{6} q_i c_i > B$, increase τ and recalculate q_i using (9.8).
  - If $\sum_{i=1}^{6} q_i c_i < B$, decrease τ and recalculate q_i using (9.8).

The values of τ can be found by bisection over the interval between the initial lower and upper bounds for τ. After each iteration, update the lower bound (in case τ should be increased) or upper bound (in case τ should be decreased), and set τ to be the mid-point between the current upper and lower bounds. Repeat until τ converges given an acceptable error margin. Since $G_i(q_i)$ in (9.8) is a cumulative distribution function, any chosen τ has to satisfy $0 \leq \frac{u_i - \tau c_i}{o_i + u_i} \leq 1$, which is equivalent to $-\frac{o_i}{c_i} \leq \tau \leq \frac{u_i}{c_i}$, for each item group. Therefore, zero and $\min\left(\frac{u_i}{c_i}\right)$ are the initial lower and upper bounds for τ.

In Table 9.1 we show an example of pricing and demand data for each group of items. We assume all demands follow a normal distribution, and there is an available budget of \$1,000,000. Note that since we are assuming a normal distribution to approximate each demand, the resulting value of q_i^* from (9.8) could be negative. However, in the derivation from (9.1) we assumed that each demand only takes positive values (the integral vanishes otherwise). For a demand distribution that can be negative valued, we assume that $G_i(0) \approx g_i(0)$, i.e., the probability of a negative demand is negligible (approximately zero). For the distributions in Table 9.1, this probability is less than 0.1% for each item group. Nevertheless, for the implementation of the solution search algorithm described above, we rewrite q_i^* from (9.8) as $q_i^* = \max\left(G_i^{-1}\left(\frac{u_i - \tau c_i}{o_i + u_i}\right), 0\right)$, that is, we constrain the solution space to non-negative values of q_i. Then, the upper bound of τ is recomputed as the minimum ratio of the underage cost and discounted price among the items for

Table 9.2 Newsvendor model results for advance purchasing quantities

	Unconstrained newsvendor	Constrained newsvendor
Cover set	6,646	1,948
Lighting set	17,716	9,938
Generators	4,174	2,833
Bottled water	52,357	31,291
Recovery set	6,775	1,137
Cleanup set	7,936	1,124
Expected advance purchasing cost	\$2,033,360	\$1,000,000

which the purchase quantity was not set to zero (a looser upper bound is of course given by $\max\left(\frac{u_i}{c_i}\right)$.

After applying (9.8) and the proposed algorithm on data shown in Table 9.1, we obtain the solutions shown in Table 9.2 for the unconstrained ($\tau = 0$) and constrained Newsvendor problem.

9.4.2 Inventory Allocation Optimization

Once the advance purchase quantities are determined, the next step is allocating the reserved inventory. As in the advance purchasing decisions, there are different trade-offs to consider. For instance, different regions are affected in different degrees. Forecasts might help to determine the potential paths and strength of the expected storms. It sounds reasonable to allocate inventory among regional DCs according to the demand expectations. However, inventory might be needed to be relocated to other DCs if demand deviates from the original allocation. In this case, holding inventory more centrally might be better.

We analyze the following problem. Inventory purchased in advance can be allocated among the regional DCs (which are assumed to be four). The vendors' prices include shipping costs; however, the company incurs additional shipping costs if the product needs to be re-allocated from one DC to another. A couple of days before a hurricane hits, there is more information about its particular path and the regions it is going to affect the most, and therefore about the demand location; also, there is some inventory of the hurricane seasonal products already allocated in the regional DCs, i.e., the products purchased in advance shipped from vendors to DCs. It is a company's policy that these products should be allocated proportionally to the expected demand; therefore, once the hurricane path is revealed and the estimated number of affected people is better known, inventory might be re-allocated to another DC to match the latest information about the demand location.

Table 9.3 Average unit transportation cost among DCs

DC/region	DC_1	DC_2	DC_3	DC_4
DC_1	\$0.00	\$0.64	\$1.46	\$2.23
DC_2	\$0.64	\$0.00	\$1.03	\$1.62
DC_3	\$1.46	\$1.03	\$0.00	\$1.74
DC_4	\$2.23	\$1.62	\$1.74	\$0.00

Table 9.4 Scenarios for demand location

Hurricane path	Probability (%)	DC_1 (%)	DC_2 (%)	DC_3 (%)	DC_4 (%)
1	30	30	40	30	0
2	20	10	25	60	5
3	10	30	0	10	60
4	40	10	10	30	50

Table 9.3 shows the (hypothetical) average unit transportation cost among every pair of DCs. For simplicity, an average transportation cost is used instead of a unit cost per item type. This is a reasonable assumption if the mix of products in the re-allocation flows does not change. Given this assumption, we can group all the group items in one product flow. Table 9.4 shows the proposed potential scenarios for demand location.

Given that the demand location scenarios are the same for all the item groups the optimal percentage of the total reserved inventory allocated in each DC does not change along the items groups. Also, this percentage does not vary with the advance purchased quantities, since it is assumed that there are not capacity constraints in the DCs. Therefore, the inventory allocation problem consists only on deciding the percentage of the inventory to allocate to each DC. A two-stage linear program (LP) is used to find the optimal inventory allocation for each regional DC.

9.4.2.1 Parameters

α_w: Probability of scenario w, $w = 1, 2, 3, 4$.
t_{ij}: Transportation cost between DC_i and DC_j, $i = 1, 2, 3, 4$; $j = 1, 2, 3, 4$. (if $i = j$, $t_{ij} = 0$).
p_{wj}: Percentage of total demand of DC_j, given scenario w.

9.4.2.2 Variables

y_i: Percentage of reserved quantity allocated to DC_i.
$x_{w,ij}$: (Percentage) flow from DC_i to DC_j for scenario w.

Table 9.5 Advance purchased (rounded) quantities allocated to DC_i for each item group

	% Allocated	Cover set	Lighting set	Generators	Bottled water	Recovery set	Cleanup set
Total	100	1,948	9,938	2,833	31,291	1,137	1,124
DC_1	10	195	994	283	3,129	114	112
DC_2	20	390	1,988	567	6,258	227	225
DC_3	30	584	2,981	850	9,387	341	337
DC_4	40	779	3,975	1,133	12,516	455	450

9.4.2.3 Optimization Model

The cost minimization model becomes:
Minimize:

$$\sum_{w}\sum_{i,j}\alpha_w t_{i,j} x_w^{i,j}. \tag{9.9}$$

Subject to:

$$\sum_{i} y_i = 1, \tag{9.10}$$

$$\sum_{j} x_w^{i,j} \leq y_i \quad \forall i = 1,2,3,4 \quad \forall w = 1,2,3,4, \tag{9.11}$$

$$\sum_{j} x_w^{i,j} \leq p_{w,j} \quad \forall j = 1,2,3,4 \quad \forall w = 1,2,3,4, \tag{9.12}$$

$$y_i, \quad x_w^{i,j} \geq 0 \quad \forall i = 1,2,3,4 \quad \forall j = 1,2,3,4 \quad \forall w = 1,2,3,4. \tag{9.13}$$

During the first stage of the problem, the allocation percentages y_i have to be specified for each DC. Then, during the second stage, inventory might be re-allocated such that the final inventory distribution is proportional to the demand of the regional DCs. Since this demand location is uncertain, and we are given only a probability α_w for each hurricane path scenario, there would be different re-allocation flows $x_w^{i,j}$ among DCs for each scenario. Nevertheless, the first stage allocation decision will affect the re-allocation transportation costs given the realization of the demand distribution scenario p_{wj}. The objective is to find the inventory allocation such that the expected re-allocation transportation cost among regional DCs is minimized. The optimization model above was programmed and run using the General Algebraic Modeling System (GAMS) (http://www.gams.com/). The optimal allocation solution resulting from the data in Tables 9.3 and 9.4 and the given constrained Newsvendor solution quantities has a re-allocation cost of $22,398. Details of the solution are shown in Table 9.5.

9.4.3 A Scenario-Based Approach

We developed a game (Ergun et al. 2008) directed to students and professionals, the users, interested in supply chain management as well as in disaster preparedness and response operations. The game setting takes place around a fictitious retailer of appliances, furniture, and general home improvement products: Big Depot. Similar to THD during the hurricane season, Big Depot aims to efficiently provide products required by the stores during the pre-strike and post-strike phases, in an accurate and timely manner. Users make recommendations about the advance purchase quantities for each group of items and the allocation of this inventory among Big Depot DCs, to minimize costs. First, users come with an approach and solutions of their own. Second, access to decision support tools is provided, and users are able to evaluate and improve their proposed solutions. There are two decision support tools: the Procurement Decision Tool (PDT) and the Allocation Decision Tool (ADT). The tools give the expected procurement and re-allocation costs given a set of decisions on the inventory advance purchasing and allocation, as well as the total procurement and re-allocation costs of scenarios of given demand size and distribution, respectively.

In the PDT, users can adjust the advance purchasing quantities while meeting the budget using rules of thumb or heuristics considering different factors related to demand variability, gaps between discounted and expedited prices, and other factors such as an item's importance. Then, users evaluate their decisions by observing their expected procurement cost and their performance under different scenarios. The expected procurement cost for the constrained Newsvendor model is the sum of the total cost of the units purchased in advance, plus the expected cost of the expedited units. Since the demand is uncertain, the number of shortage units is unknown, so we compute an expected value for these shortage units. Since the demand is assumed to have a normal distribution, the expected shortage is computed using the Standard Normal Loss Function which is commonly reported in tables or can be calculated as follows:

Let:

$$k_i = \frac{q_i - \mu_i}{\sigma_i}. \tag{9.14}$$

Then, the expected number of shortage units is given by:

$$L_i(k_i) = (g_i(k_i) - k_i(1 - G_i(k_i)))\sigma_i \tag{9.15}$$

where g_i is the probability density function of item group i, G_i is the cumulative distribution function, and μ_i and σ_i are the mean and standard deviation, respectively. There are several heuristics that the users might try. For example, users might propose ordering the average demand of the products, but in this case the budget is violated and the quantities should be adjusted. One option is to adjust these quantities by a percentage until the budget constraint is met. However, this heuristic does not consider variability or the price gap between discount and expedited units. Another simple heuristic could be to adjust the 95th percentile quantity until the budget is met, so that the worst-case scenario of a particular product is someway captured. Users familiar with the Newsvendor Model might derive the unconstrained Newsvendor quantities first, and then adjust these quantities to meet the budget. By using a scenario-based analysis with the aid of a tool like PDT, the user can also consider objectives other than minimizing cost, can observe different trade-offs, and can evaluate the robustness of the proposed solutions.

In the ADT, users enter the percentage of the inventory purchased in advance to allocate in each of the DCs. Once the initial allocation of the advanced purchased quantities is defined, a simple heuristic could be used to find the optimal re-allocation flows once a given demand distribution scenario is given. The heuristic works as follows:

For each scenario w:

- Sort the unit transportation costs from cheapest to most expensive.
- Compute the product demand surplus or deficit for each regional DC according to the demand scenario.
- Select the cheapest route DC_i to DC_j such that there is a demand surplus in DC_i and a demand deficit in DC_j.
- Take $x_w^{i,j} = \min$ (surplus, deficit).
- Adjust the demand surplus or deficit of DC_i and DC_j subtracting $x_w^{i,j}$.
- Select the next cheapest route DC_i to DC_j such that there is a demand surplus in DC_i and a demand deficit in DC_j.
- Repeat *steps 4–6* until there is no demand surplus or deficit left in any DC.

The only decision required is the initial allocation of inventory, and the optimal re-allocation decisions can be obtained by applying the algorithm above. Similarly to the PDT, the expected re-allocation transportation cost is obtained with the cost of the different scenarios (storm trajectories) by computing a pondered average cost using the probability of each of them. While the two-stage LP above solves for the minimum re-allocation cost, using a scenario-based approach allows the user to incorporate various objectives, such as to minimize the cost of the worst-case scenario. Heuristics can include, for example, allocating the inventory according to the most likely path or allocating the inventory according to a weighted average, i.e., the percentage of inventory allocated to each DC is the result of weighting the percentage of demand assigned to each DC on each storm path, using the probability of such scenario. This heuristic would be the optimal if all transportation costs among the DCs were the same. To take into account the different transportation costs between DCs, another heuristic could be used to consolidate inventory at the most centrally located DCs.

9.5 Conclusions

Many organizations in the private sector are now taking an active role in the disaster response operations of their communities. There are many lessons that other disaster responders, such as government agencies and NGOs, can learn from the private sector, including, for instance, the empowerment of local management and the emphasis on coordination among the different functional areas and management levels. Moreover, in order to succeed as first responders, organizations need a responsive supply chain. Procurement and inventory allocation under uncertainty are just two examples of the decisions that a supply chain planner has to make in a disaster response setting. These decisions are not unique to the disaster setting, and most of the traditional OR approaches and techniques (optimization, simulation, etc.) applied in traditional supply chain management are applicable with the appropriate modifications to account for the specific disaster setting. For example, the objective function of a traditional logistics model may change when a company might be more interested in delivering an adequate response than being profitable. Also, decisions must take into account the very uncertain setting of a disaster and be adequately robust to different potential outcomes. For example, while deciding the best transportation mode or the best transportation route, the decision maker should consider potentially damaged road infrastructure, airport conditions, etc. Disaster response poses additional challenges beyond traditional supply chain planning, which means additional opportunities on how operations researchers and supply chain management professionals could apply their knowledge and skills to positively impact lives.

Acknowledgments The authors gratefully acknowledge The Home Depot personnel for sharing generously their time, experience, and process documentation for writing of the case on which this chapter is based. Support for this work was provided by the Georgia Institute of Technology Focused Research Program on Humanitarian Response, the Harold R. and Mary Anne Nash Endowment, and National Science Foundation Grant SBE-0624269. Any opinions, findings, conclusions, or recommendations expressed in this material are of the authors and do not necessarily reflect the views of the National Science Foundation.

Appendix

The Big Depot Hurricane Planning Game (Ergun et al. 2010c), including the decision support tools described above, as well as the Humanitarian Response Planning at The Home Depot case study (Ergun et al. 2010b) as a stand-alone document are both available at http://humanitarian.gatech.edu. Teaching notes, which include recommendations of the use of the case and game in the classroom, can be requested through the Web site or by emailing the authors. The case and the game could be used together or separately. However, it is recommended that the users discuss the case before working on the game. This would help them to form a better background and understand the complexities behind a disaster response planning process, characterized by the several interactions of the different functional areas, and the many decisions faced before, during, and after a disaster strikes. Then, the game would help the users to experience firsthand the uncertainty faced when dealing with supply chain decisions in situations such as disaster response.

References

Altay N, Green WG (2006) OR/MS research in disaster operations management. Eur J Oper Res 175(1):475–493

Barbaro M (2005) Wal-Mart at Forefront of Hurricane Relief. Washington. Washington Post. http://www.washingtonpost.com/wp-dyn/content/article/2005/09/05/AR2005090501598.html. Accessed 14 Aug 2011

Ergun O, Heier Stamm J, Keskinocak P, Swann J (2010a) Waffle House restaurants hurricane response: a case study. Int J Prod Econ 126:111–120

Ergun O, Drake M, Kerl P, Keskinocak P, Swann J, Villarreal M (2010b) Humanitarian response planning at The Home Depot. http://humanitarian.gatech.edu. Accessed 1 Oct 2011

Ergun O, Karakus G, Keskinocak P, Swann J, Villarreal M (2010c) The big depot hurricane planning game. http://humanitarian.gatech.edu. Accessed 1 Oct 2011

Ergun O, Karakus G, Keskinocak P, Swann J, Villarreal M (2011) Operations research to improve disaster supply chain management, In: Cochran JJ, Cox LA, Keskinocak P, Kharoufeh JP, Smith JC (eds) Wiley Encyclopedia of Operations Research and Management Science, Wiley, Hoboken, New Jersey

Hayes E (2005) What can Wal-Mart teach FEMA about disaster response? ABC News. http://abcnews.go.com/WNT/HurricaneRita/story?id=1171087&page=1. Accessed 14 Aug 2011

http://corporate.homedepot.com/en_US/Corporate/Investor_Relations/Annual_Reports/annual1992.pdf. Accessed 18 May 2011

Huffman M (2008) Big Box Retailers, not FEMA, may be first line of defense. Consumer affairs. http://www.consumeraffairs.com/news04/2008/04/hurricane_responders.html. Accessed 14 Aug 2011

King M (2011) Waffle House Index helps to weather disasters. 11Alive. http://www.11alive.com/news/article/197574/40/Waffle-House-Index-helps-to-weather-disasters. Accessed 14 Aug 2011

Lohr S (1992) Lessons from a Hurricane: it pays not to Gouge. New York Times. http://query.nytimes.com/gst/fullpage.html?res=9E0CE4D81431F931A1575AC0A964958260&sec=&spon=&pagewanted=print. Accessed 18 May 2011

The Home Depot (1992) Annual Report. http://corporate.homedepot.com/en_US/Corporate/Investor_Relations/Annual_Reports/annual1992.pdf. Accessed 18 May 2011

The Home Depot Corporate Website. http://corporate.homedepot.com/wps/portal/. Accessed 18 May 2011

Index

J.W. Herrmann (ed.), *Handbook of Operations Research for Homeland Security*, International Series in Operations Research & Management Science 183, DOI 10.1007/978-1-4614-5278-2,

Printed by Publishers' Graphics LLC
MO20121115.11.35.216